A History of the 17th Lancers

A History Of the 17th Lancers

(DUKE OF CAMBRIDGE'S OWN)

BY

HON. J. W. FORTESCUE

John Hale

First Colonel of the 17th Light Dragoons.

To the Memory

OF

MAJOR-GENERAL JAMES WOLFE

WHO FELL GLORIOUSLY IN THE MOMENT OF VICTORY

ON THE PLAINS OF ABRAHAM BEFORE QUEBEC

13TH SEPTEMBER 1759

THIS HISTORY

OF THE REGIMENT RAISED IN HIS HONOUR

BY HIS COMRADE IN ARMS

JOHN HALE

IS PROUDLY AND REVERENTLY INSCRIBED

Farrier. Officer. Trumpeter.

1763.

Marching Order. Field-day Order. Review Order.

PRIVATES, 1784 - 1810.

Field-day Order. Review Order.

OFFICERS, 1810 – 1813.

Watering Order. Review Order. Marching Order.

PRIVATES, 1810 – 1813.

G. Salisbury.

OFFICER, Review Order. PRIVATE, Field-day Order. CORPORAL, Marching Order.

1814.

G. Salisbury.

OFFICER, Review Order. PRIVATE, Field-day Order. OFFICER, Stable Dress.

1817 – 1823

G. Salisbury, 1832.

OFFICERS, 1824.

G. Salisbury.

Marching Order. Review Order.

G. Salisbury, 1832.

OFFICERS, 1829

G. Salisbury.

PRIVATE, Review Order. OFFICER, Marching Order. PRIVATE, Marching Order.

1829 – 1832.

G. Salisbury.

Review Order. Marching Order.

OFFICERS. 1832 – 1841.

INDIA 1858.

1894.

Preface

THIS history has been compiled at the request of the Colonel and Officers of the Seventeenth Lancers.

The materials in possession of the Regiment are unfortunately very scanty, being in fact little more than the manuscript of the short, and not very accurate summary drawn up nearly sixty years ago for Cannon's *Historical Records of the British Army*. The loss of the regimental papers by shipwreck in 1797 accounts for the absence of all documents previous to that year, as also, I take it, for the neglect to preserve any sufficient records during many subsequent decades. I have therefore been forced to seek information almost exclusively from external sources.

The material for the first three chapters has been gathered in part from original documents preserved in the Record Office,—Minutes of the Board of General Officers, Muster-Rolls, Pay-sheets, Inspection Returns, Marching Orders, and the like; in part from a mass of old drill-books, printed Standing Orders, and military treatises, French and English, in the British Museum. The most important of these latter are Dalrymple's *Military Essay*, Bland's *Military Discipline*, and, above all, Hinde's *Discipline of the Light Horse* (1778).

For the American War I have relied principally on the

original despatches and papers, numerous enough, in the Record Office, Tarleton's *Memoirs*, and Stedman's *History of the American War*,—the last named being especially valuable for the excellence of its maps and plans. I have also, setting aside minor works, derived much information from the two volumes of the *Clinton-Cornwallis Controversy* compiled by Mr. B. Stevenson; and from Clinton's original pamphlets, with manuscript additions in his own hand, which are preserved in the library at Dropmore.

For the campaigns in the West Indies the original despatches in the Record Office have afforded most material, supplemented by a certain number of small pamphlets in the British Museum. The Maroon War is treated with great fulness by Dallas in his *History of the Maroons;* and there is matter also in Bridges' *Annals of Jamaica*, and the works of Bryan Edwards. The original despatches are, however, indispensable to a right understanding of the war. Unfortunately the despatches that relate to St. Domingo are not to be found at the Record Office, so that I have been compelled to fall back on the few that are published in the *London Gazette.* Nor could I find any documents relating to the return of the Regiment from the West Indies, which has forced me unwillingly to accept the bald statement in Cannon's records.

The raid on Ostend and the expedition to La Plata have been related mainly from the accounts in the original despatches, and from the reports of the courts-martial on General Whitelocke and Sir Home Popham. There is much interesting information as to South America,—original memoranda by Miranda, Popham, Sir Arthur Wellesley (the Duke of Wellington) and other documents—preserved among the manuscripts at Dropmore.

Preface

The dearth of original documents both at the Record Office and the India Office has seriously hampered me in tracing the history of the Regiment during its first sojourn in India and through the Pindari War. I have, however, to thank the officials of the Record Department of the India Office for the ready courtesy with which they disinterred every paper, in print or manuscript, which could be of service to me.

Respecting the Crimea and the Indian Mutiny I have received (setting aside the standard histories) much help from former officers, notably Sir Robert White, Sir William Gordon, and Sir Drury Lowe, but especially from Sir Evelyn Wood, who kindly found time, amid all the pressure of his official duties, to give me many interesting particulars respecting the chase of Tantia Topee. Above all I have to thank Colonel John Brown for information and assistance on a hundred points. His long experience and his accurate memory, quickened but not clouded by his intense attachment to his old regiment, have been of the greatest value to me.

My thanks are also due to the officials of the Record Department of the War Office, and to Mr. S. M. Milne of Calverley House, Leeds, for help on divers minute but troublesome points, and to Captain Anstruther of the Seventeenth Lancers for constant information and advice. Lastly, and principally, let me express my deep obligations to Mr. Hubert Hall for his unwearied courtesy and invaluable guidance through the paper labyrinth of the Record Office, and to Mr. G. K. Fortescue, the Superintendent of the Reading-Room at the British Museum, for help rendered twice inestimable by the kindness wherewith it was bestowed.

The first and two last of the coloured plates in this book have been taken from original drawings by Mr J. P. Beadle. The remainder are from old drawings, by one G. Salisbury, in the possession of the regiment. They have been deliberately chosen as giving, on the whole, a more faithful presentment of the old and extinct British soldier than could easily be obtained at the present day, while their defects are of the obvious kind that disarm criticism. The portrait of Colonel John Hale is from an engraving after a portrait by Sir Joshua Reynolds, the original of which is still in possession of his lineal descendant in America. That of Lord Bingham is after a portrait kindly placed at the disposal of the Regiment by his son, the present Earl of Lucan. Those of the Duke of Cambridge and of Sir Drury Lowe are from photographs.

May, 1895.

Contents

CHAP.		PAGE
1.	The Rise of the 17th Light Dragoons, 1759 . .	1
2.	The Making of the 17th Light Dragoons . . .	10
3.	Reforms after the Peace of Paris, 1763-1774 . .	20
4.	The American War—1st Stage—The Northern Campaign, 1775-1780	31
5.	The American War—2nd Stage—The Southern Campaign, 1780-1782	49
6.	Return of the 17th from America, 1783—Ireland, 1793—Embarkation for the West Indies, 1795 . .	65
7.	The Maroon War in Jamaica, 1795 . . .	73
8.	Grenada and St. Domingo, 1796	87
9.	Ostend—La Plata, 1797-1807	96
10.	First Sojourn of the 17th in India, 1808-1823—The Pindari War	110
11.	Home Service, 1823-1854	121
12.	The Crimea, 1854-1856	128
13.	Central India, 1858-1859	144
14.	Peace Service in India and England, 1859-1879 . .	166
15.	The Zulu War—Peace Service in India and at Home, 1879-1894	174

Appendix

PAGE

A. A List of the Officers of the 17th Light Dragoons, Lancers 181

B. Quarters and Movements of the 17th Lancers since their Foundation 236

C. Pay of all Ranks of a Light Dragoon Regiment, 1764 . 241

D. Horse Furniture and Accoutrements of a Light Dragoon, 1759 243

E. Clothing, etc. of a Light Dragoon, 1764 . . . 244

F. Evolutions required at the Inspection of a Regiment, 1759 . 245

List of Illustrations

Lieutenant-Colonel John Hale

H.R.H. The Duke of Cambridge, K.G., Colonel-in-Chief 17th Lancers

Seventeenth Light Dragoons, 1764

Privates, 1784-1810

Officers, 1810-1813

Privates, 1810-1813

Officer, Corporal, and Privates, 1814

Officers and Private, 1817-1823

Officers, 1824

Privates, 1824-1829

George, Lord Bingham

Officers, 1829

Officer and Privates, 1829-1832

Officers, 1832-1841

Central India, 1858, 1859

Lieutenant-General Sir Drury Curzon Drury Lowe, K.C.B.

Seventeenth Lancers, 1895

W. & D. Downey Photo. Walker & Boutall Ph. Sc.

H.R.H. The Duke of Cambridge, K.G.
Colonel-in-chief 17th Lancers, 1876.

CHAPTER I

THE RISE OF THE 17TH LIGHT DRAGOONS, 1759

THE British Cavalry Soldier and the British Cavalry Regiment, such as we now know them, may be said to date from 1645, that being the year in which the Parliamentary Army, then engaged in fighting against King Charles the First, was finally remodelled. At the outbreak of the war the Parliamentary cavalry was organised in seventy-five troops of horse and five of dragoons: the Captain of the 67th troop of horse was Oliver Cromwell. In the winter of 1642-43 Captain Cromwell was promoted to be Colonel, and entrusted with the task of raising a regiment of horse. This duty he fulfilled after a fashion peculiarly his own. Hitherto the Parliamentary horse had been little better than a lot of half-trained yeomen: Colonel Cromwell took the trouble to make his men into disciplined cavalry soldiers. Moreover, he raised not one regiment, but two, which soon made a mark by their superior discipline and efficiency, and finally at the battle of Marston Moor defeated the hitherto invincible cavalry of the Royalists. After that battle Prince Rupert, the Royalist cavalry leader, gave Colonel Cromwell the nickname of Ironside; the name was passed on to his regiments, which grew to be known no longer as Cromwell's, but as Ironside's.

1645.

In 1645, when the army was remodelled, these two famous regiments were taken as the pattern for the English cavalry; and having been blent into one, appear at the head of the list as Sir Thomas Fairfax's Regiment of Horse. Fairfax was General-in-Chief, and his appointment to the colonelcy was of course a

1645. compliment to the regiment. Besides Fairfax's there were ten other regiments of horse, each consisting of six troops of 100 men apiece, including three corporals and two trumpeters. As the field-officers in those days had each a troop of his own, the full establishment of the regiments was 1 colonel, 1 major, 4 captains, 6 lieutenants, 6 cornets, 6 quartermasters. Such was the origin of the British Cavalry Regiment.

The troopers, like every other man in this remodelled army, wore scarlet coats faced with their Colonel's colours—blue in the case of Fairfax. They were equipped with an iron cuirass and an iron helmet, armed with a brace of pistols and a long straight sword, and mounted on horses mostly under fifteen hands in height. For drill in the field they were formed in five ranks, with six feet (one horse's length in those days), both of interval and distance, between ranks and files, so that the whole troop could take ground to flanks or rear by the simple words, "To your right (or left) turn;" "To your right (or left) about turn." Thus, as a rule, every horse turned on his own ground, and the troop was rarely wheeled entire: if the latter course were necessary, ranks and files were closed up till the men stood knee to knee, and the horses nose to croup. This formation deservedly bore the name of "close order." For increasing the front the order was, "To the right (or left) double your ranks," which brought the men of the second and fourth ranks into the intervals of the first and third, leaving the fifth rank untouched. To diminish the front the order was: "To the right (or left) double your files," which doubled the depth of the files from five to ten in the same way as infantry files are now doubled at the word, "Form fours."

The principal weapons of the cavalry soldiers were his firearms, generally pistols, but sometimes a carbine. The lance, which had formerly been the favourite weapon, at Crecy for instance, was utterly out of fashion in Cromwell's time, and never employed when any other arm was procurable. Firearms were the rage of the day, and governed the whole system of cavalry attack. Thus in action the front rank fired its two pistols, and filed away to load again in the rear, while the second and third ranks came up and did likewise. If the word were given to charge, the men advanced to the charge pistol in hand, fired, threw it in the enemy's face, and then fell in with the sword. But though there was a very elaborate exercise for carbine and pistol, there was no such thing as sword exercise.

1645.

Moreover, though the whole system of drill was difficult, and required perfection of training in horse and man, yet there was no such thing as a regular riding-school. If a troop horse was a kicker a bell was placed on his crupper to warn men to keep clear of his heels. If he were a jibber the following were the instructions given for his cure :—

" If your horse be resty so as he cannot be put forwards then let one take a cat tied by the tail to a long pole, and when he [the horse] goes backward, thrust the cat within his tail where she may claw him, and forget not to threaten your horse with a terrible noise. Or otherwise, take a hedgehog and tie him strait by one of his feet to the horse's tail, so that he [the hedgehog] may squeal and prick him."

For the rest, certain peculiarities should be noted which distinguish cavalry from infantry. In the first place, though every troop and every company had a standard of its own, such standard was called in the cavalry a Cornet, and in the infantry an Ensign, and gave in each case its name to the junior subaltern whose duty it was to carry it. In the second place there were no sergeants in old days except in the infantry, the non-commissioned officers of cavalry being corporals only. In the third place, the use of a wind instrument for making signals was confined to the cavalry, which used the trumpet ; the infantry as yet had no bugle, but only the drum. There were originally but six trumpet-calls, all known by foreign names ; of which names one (*Butte sella* or *Boute selle*) still survives in the corrupted form, " Boots and saddles."

How then have these minor distinctions which formerly separated cavalry from infantry so utterly disappeared ? Through

1645. what channel did the two branches of the service contrive to meet? The answer is, through the dragoons. Dragoons were originally mounted infantry pure and simple. Those of the Army of 1645 were organised in ten companies, each 100 men strong. They were armed like infantry and drilled like infantry; they followed an ensign and not a cornet; they obeyed, not a trumpet, but a drum. True, they were mounted, but on inferior horses, and for the object of swifter mobility only; for they always fought on foot, dismounting nine men out of ten for action, and linking the horses by the rude process of throwing each animal's bridle over the head of the horse standing next to it in the ranks. Such were the two branches of the mounted service in the first British Army.

1745. A century passes, and we find Great Britain again torn by internal strife in the shape of the Scotch rebellion. Glancing at the list of the British cavalry regiments at this period we find them still divided into horse and dragoons; but the dragoons are in decided preponderance, and both branches unmistakably "heavy." A patriotic Englishman, the Duke of Kingston, observing this latter failing, raised a regiment of Light Horse (the first ever seen in England) at his own expense, in imitation of the Hussars of foreign countries. Thus the Civil War of 1745 called into existence the only arm of the military service which had been left uncreate by the great rebellion of 1642-48. Before leaving this Scotch rebellion of 1745, let us remark that there took part in the suppression thereof a young ensign of the 47th Foot, named John Hale—a mere boy of seventeen, it is true, but a promising officer, of whom we shall hear more.

The Scotch rebellion over, the Duke of Kingston's Light Horse were disbanded and re-established forthwith as the Duke of Cumberland's own, a delicate compliment to their distinguished service. As such they fought in Flanders in 1747, but were finally disbanded in the following year. For seven years after the British Army possessed no Light Cavalry, until at the end of 1755 a single troop of Light Dragoons—3 officers and 65 men strong—was added to each of the eleven cavalry regiments on the British establishment, viz., the 1st, 2nd, and 3rd Dragoon Guards, and the 1st, 2nd, 3rd, 4th, 6th, 7th, 10th, and 11th Dragoons. These light dragoons were armed with carbine and bayonet and a single pistol, the second holster being filled (sufficiently filled, one must conclude) with an axe, a hedging-bill, and a spade. Their shoulder-belts were provided with a swivel to which the carbine could be sprung; for these light troops were expected to do a deal of firing from the saddle. Their main distinction of dress was that they wore not hats like the rest of the army, but helmets —helmets of strong black jacked leather with bars down the sides and a brass comb on the top. The front of the helmet was red, ornamented with the royal cypher and the regimental number in brass; and at the back of the comb was a tuft of horse-hair, half coloured red for the King, and half of the hue of the regimental facings for the regiment. The Light Dragoon-horse, we learn, was of the "nag or hunter kind," standing from 14.3 to 15.1, for he was not expected to carry so heavy a man nor such cumbrous saddlery as the Heavy Dragoon-horse. Of this latter we can only say that he was a most ponderous animal, with a character of his own, known as the "true dragoon mould, short-backed, well-coupled, buttocked, quartered, forehanded, and limbed,"—all of which qualities had to be purchased for twenty guineas. At this time, and until 1764, all troop horses were docked so short that they can hardly be said to have kept any tail at all. 1745.

In the year 1758 nine of these eleven light troops took part in an expedition to the coast of France, England having two years before allied herself with Prussia against France for the great struggle known as the Seven Years' War. So eminent was the service which they rendered, that in March 1759, King George II. decided to raise an entire regiment of Light Dragoons. On the 10th of March, accordingly, the first regiment was raised by General Elliott and numbered the 15th. The Major of this regiment, whom we shall meet again as Brigadier of cavalry in America, was William Erskine. On the 4th August another 1759.

1759. regiment of Light Dragoons was raised by Colonel Burgoyne, and numbered the 16th. We shall see the 16th distinguished and Burgoyne disgraced before twenty years are past.

And while these two first Light Dragoon regiments are a-forming, let us glance across the water to Canada, where English troops are fighting the French, and seem likely to take the country from them. Among other regiments the 47th Foot is there, commanded (since March 1758) by Colonel John Hale, the man whom we saw fighting in Scotland as an ensign fourteen years ago. Within the past year he has served with credit under General Amherst at the capture of Cape Breton and Louisburg, and in these days of August, while Burgoyne is raising his regiment, he is before Quebec with General Wolfe. Three months more pass away, and on the 13th of October Colonel John Hale suddenly arrives in London. He is the bearer of despatches which are to set all England aflame with pride and sorrow; for on the 13th of September was fought the battle on the plains of Abraham which decided the capture of Quebec and the conquest of Canada. General Wolfe fell at the head of the 28th Regiment in the moment of victory; and Colonel Hale, who took a brilliant share in the action at the head of the 47th, has been selected to carry the great news to the King. Colonel Hale was well received; the better for that Wolfe's last despatches, written but four days before the battle, had been marked by a tone of deep despondency; and, we cannot doubt, began to wonder what would be his reward. He did not wonder for long.

Very shortly after Hale's arrival the King reviewed the 15th Light Dragoons, and was so well pleased with their appearance that he resolved to raise five more such regiments, to be numbered the 17th to the 21st.

The raising of the first of these regiments, that now known to us as the Seventeenth Lancers, was intrusted to Colonel John Hale, who received his commission for the purpose on the 7th

November. For the time, however, the regiment was known as the Eighteenth, for what reason it is a little difficult to understand; since the apology for a corps which received the number Seventeen was not raised until a full month later (December 19th). As we shall presently see, this matter of the number appears to have caused some heartburning, until Lord Aberdour's corps, which had usurped the rank of Seventeenth, was finally disbanded, and thus yielded to Hale's its proper precedence. 1759.

On the very day when Colonel Hale's commission was signed, which we may call the birthday of the Seventeenth Lancers, the Board of General Officers was summoned to decide how the new regiment should be dressed. As to the colour of the coat there could be no doubt, scarlet being the rule for all regiments. For the facings white was the colour chosen, and for the lace white with a black edge, the black being a sign of mourning for the death of Wolfe. But the principal distinction of the new regiment was the badge, chosen by Colonel Hale and approved by the King, of the Death's Head and the motto "Or Glory,"—the significance of which lies not so much in claptrap sentiment, as in the fact that it is, as it were, a perpetual commemoration of the death of Wolfe. It is difficult for us to realise, after the lapse of nearly a century and a half, how powerfully the story of that death seized at the time upon the minds of men. 7th Nov.

Two days after the settlement of the dress, a warrant was issued for the arming of Colonel Hale's Light Dragoons; and this, being the earliest document relating to the regiment that I have been able to discover, is here given entire:—

George R.

Whereas we have thought fit to order a Regiment of Light Dragoons to be raised and to be commanded by our trusty and well-beloved Lieutenant-Colonel John Hale, which Regiment is to consist of Four troops, of 3 sergeants, 3 corporals, 2 drummers, and 67 private men in each troop, besides commission officers, Our will and pleasure is, that out of the stores remaining within the Office of our Ordnance under your charge you cause 300 pairs of pistols, 292 carbines, 292 cartouche boxes, and 8 drums, to be issued and delivered to the said Lieutenant-Colonel John Hale, or to such

1759. person as he shall appoint to receive the same, taking his indent as usual, and you are to insert the expense thereof in your next estimate to be laid before Parliament. And for so doing this shall be as well to you as to all other our officers and ministers herein concerned a sufficient Warrant.

Given at our Court at St. James' the 9th day of November 1759, in the 33rd year of our reign.

To our trusty and well-beloved Cousin and Councillor John Viscount Ligonier, Master-General of our Ordnance.

These preliminaries of clothing and armament being settled, Colonel Hale's next duty was to raise the men. Being a Hertfordshire man, the son of Sir Bernard Hale of Kings Walden, he naturally betook himself to his native county to raise recruits among his own people. The first troop was raised by Captain Franklin Kirby, Lieutenant, 5th Foot; the second by Captain Samuel Birch, Lieutenant, 11th Dragoons; the third by Captain Martin Basil, Lieutenant, 15th Light Dragoons; and the fourth by Captain Edward Lascelles, Cornet, Royal Horse Guards. If it be asked what stamp of man was preferred for the Light Dragoons, we are able to answer that the recruits were required to be "light and straight, and by no means gummy," not under 5 feet 5½ inches, and not over 5 feet 9 inches in height. The bounty usually offered (but varied at the Colonel's discretion) was three guineas, or as much less as a recruit could be persuaded to accept.

Whether from exceptional liberality on the part of Colonel Hale, or from an extraordinary abundance of light, straight, and by no means gummy men in Hertfordshire at that period, the regiment was recruited up to its establishment, we are told, within the space of seventeen days. Early in December it made rendezvous at Watford and Rickmansworth, whence it marched to Warwick and Stratford-on-Avon, and thence a fortnight later to Coventry. Meanwhile orders had already been given (10th December) that its establishment should be augmented by two more troops of the same strength as the original four; and little more than a month later came a second order to increase each

December.

1760. 28th Jan.

of the existing troops still further by the addition of a sergeant, a corporal, and 36 privates. Thus the regiment, increased almost as soon as raised from 300 to 450 men, and within a few weeks again strengthened by one-half, may be said to have begun life with an establishment of 678 rank and file. To them we must add a list of the original officers :— 1760.

Lieutenant-Colonel Commandant.—John Hale, 7th November 1759.

Major.—John Blaquiere, 7th November 1759.

Captains.		Lieutenants.		Cornets.	
Franklin Kirby .	4th Nov.	Thomas Lee .	4th Nov.	Robert Archdall .	4th Nov.
Samuel Birch .	5th "	William Green .	5th "	Henry Bishop .	5th "
Martin Basil .	6th "	Joseph Hall .	6th "	Joseph Stopford .	6th "
Edward Lascelles .	7th "	Henry Wallop .	7th "	Henry Crofton .	7th "
John Burton .	7th "	Henry Cope .	7th "	Joseph Moxham .	7th "
Samuel Townshend	8th "	Yelverton Peyton .	8th "	Daniel Brown .	8th "

Adjutant.—Richard Westbury. *Surgeon.*—John Francis.

CHAPTER II

THE MAKING OF THE 17TH LIGHT DRAGOONS

1760. DETAILS of the regiment's stay at Coventry are wanting, the only discoverable fact being that, in obedience to orders from headquarters, it was carefully moved out of the town for three days in August during the race-meeting. But as these first six months must have been devoted to the making of the raw recruits into soldiers, we may endeavour, with what scanty material we can command, to form some idea of the process. First, we must premise that with the last order for the augmentation of establishment was issued a warrant for the supply of the regiment with bayonets, which at that time formed an essential part of a dragoon's equipment. Swords, it may be remarked, were provided, not by the Board of Ordnance, but by the Colonel. It is worth while to note in passing how strong the traditions of 1645 still remain in the dragoons. The junior subaltern is indeed no longer called an ensign, but a cornet; but the regiment is still ruled by the infantry drum instead of the cavalry trumpet.

Let us therefore begin with the men; and as we have already seen what manner of men they were, physically considered, let us first note how they were dressed. Strictly speaking, it was not until 1764 that the Light Dragoon regiments received their distinct dress regulations; but the alterations then made were so slight that we may fairly take the dress of 1764 as the dress of 1760. To begin with, every man was supplied by the Colonel, by contract, with coat, waistcoat, breeches, and cloak. The coat, of course, was of scarlet, full and long in the skirt, but whether lapelled or not before 1763 it is difficult to say. Lapels meant a good deal in those days; the coats of Horse being lapelled to the skirt, those of Dragoon Guards lapelled to the waist, while those of Dragoons were double-breasted and had no lapels at all. The Light Dragoons being a novelty, it is difficult to say how they were distinguished in this respect, but probably in 1760 (and certainly in 1763) their coats were lapelled to the waist with the colour of the regimental facing, the lapels being three inches broad, with plain white buttons disposed thereon in pairs.

1760.

The waistcoat was of the colour of the regimental facing—white, of course, for the Seventeenth; and the breeches likewise. The cloaks were scarlet, with capes of the colour of the facing. In fact, it may be said once for all that everything white in the uniform of the Seventeenth owes its hue to the colour of the regimental facing.

Over and above these articles the Light Dragoon received a pair of high knee-boots, a pair of boot-stockings, a pair of gloves, a comb, a watering or forage cap, a helmet, and a stable frock. Pleased as the recruit must have been to find himself in possession of smart clothes, it must have been a little discouraging for him to learn that his coat, waistcoat, and breeches were to last him for two, and his helmet, boots, and cloak for four years. But this was not all. He was required to supply out of an annual wage of £13 : 14 : 10 the following articles at his own expense:—

4 shirts at 6s. 10d.	£1	7	4
4 pairs stockings at 2s. 10d.	0	11	4
2 pairs shoes at 6s.	0	12	0
A black stock	0	0	8
Stock-buckle	0	0	6
1 pair leather breeches	1	5	0
1 pair knee-buckles	0	0	8
2 pairs short black gaiters	0	7	4
1 black ball (the old substitute for blacking)	0	1	0
3 shoe-brushes	0	1	3
	£4	7	1

1760. Nor was even this all, for we find (though without mention of their price) that a pair of checked sleeves for every man, and a powder bag with two puffs for every two men had likewise to be supplied from the same slender pittance.

Turning next from the man himself to his horse, his arms, and accoutrements, we discover yet further charges against his purse, thus—

	£	s.	d.
Horse-picker and turnscrew	£0	0	2
Worm and oil-bottle	0	0	3½
Goatskin holster tops	0	1	6
Curry-comb and brush . . .	0	2	3
Mane comb and sponge . . .	0	0	8
Horse-cloth	0	4	9
Snaffle watering bridle . . .	0	2	0
	£0	11	7½

Also a pair of saddle-bags, a turn-key, and an awl.

All these various items were paid for, "according to King's regulation and custom," out of the soldier's "arrears and grass money." For his pay was made up of three items—

	£	s.	d.	
"Subsistence" (5d. a day nominal) .	£9	2	0	per annum.
"Arrears" (2d. a day nominal) . .	3	1	0	"
"Grass money" . . .	1	11	10	"
	£13	14	10	"

We must therefore infer that his "subsistence" could not be stopped for his "necessaries" (as the various items enumerated above are termed); but none the less twopence out of the daily stipend was stopped for his food, while His Majesty the King deducted for his royal use a shilling in the pound from the pay of every soul in the army. Small wonder that heavy bounty-money was needed to persuade men to enlist.

What manner of instruction the recruit received on his first appearance it is a little difficult to state positively, though it is still possible to form a dim conception thereof. The first thing that he was taught, apparently, was the manual and firing exercise, 1760. of which we are fortunately able to speak with some confidence. As it contains some eighty-eight words of command, we may safely infer that by the time a recruit had mastered it he must have been pretty well disciplined. The minuteness of the exercise and the extraordinary number of the motions sufficiently show that it counted for a great deal. "The first motion of every word of command is to be performed immediately after it is given; but before you proceed to any of the other motions you must tell one, two, pretty slow, by making a stop between the words, and in pronouncing the word *two*, the motion is to be performed." In those days the word "smart" was just coming into use, but "brisk" is the more common substitute. Let us picture the squad of recruits with their carbines, in their stable frocks, white breeches, and short black gaiters, and listen to the instructions which the corporal is giving them:—

"Now on the word *Shut your pans*, let fall the primer and take hold of the steel with your right hand, placing the thumb in the upper part, and the two forefingers on the lower. Tell *one*, *two*, and shut the pan; tell *one*, *two*, and seize the carbine behind the lock with the right hand; then tell *one*, *two*, and bring your carbine briskly to the recover. Wait for the word. Shut your —pans, one—two, one—two, one—two."

There is no need to go further through the weary iteration of "Join your right hand to your carbine," "Poise your carbine," "Join your left hand to your carbine," whereby the recruit learned the difference between his right hand and his left. Suffice it that the manual and firing exercise contain the only detailed instruction for the original Light Dragoon that is now discoverable. "Setting-up" drill there was apparently none, sword exercise there was none, riding-school, as we now understand it, there was none, though there was a riding-master. A "ride" appears to have comprised at most twelve men, who moved in a circle round the riding-master and received his teaching as best they could. But it must not be inferred on that account that the men could not

1760. ride; on the contrary the Light Dragoons seem to have particularly excelled in horsemanship. Passaging, reining back, and other movements which call for careful training of man and horse, were far more extensively used for purposes of manœuvre than at present. Moreover, every man was taught to fire from on horseback, even at the gallop; and as the Light Dragoons received an extra allowance of ammunition for ball practice, it is reasonable to conclude that they spent a good deal of their time at the butts, both mounted and dismounted.

As to the ordinary routine life of the cavalry barrack, it is only possible to obtain a slight glimpse thereof from scattered notices. Each troop was divided into three squads with a corporal and a sergeant at the head of each. Each squad formed a mess; and it is laid down as the duty of the sergeants and corporals to see that the men "boil the pot every day and feed wholesome and clean." The barrack-rooms and billets must have been pretty well filled, for every scrap of a man's equipment, including his saddle and saddle-furniture, was hung up therein according to the position of his bed. As every bed contained at least two men, there must have been some tight packing. It is a relief to find that the men could obtain a clean pair of sheets every thirty days, provided that they returned the foul pair and paid three halfpence for the washing.

The fixed hours laid down in the standing orders of the Light Dragoons of 14th May 1760 are as follows:—

The drum beat for—

Réveille from Ladyday to Michaelmas			5.30 A.M.	Rest of year	6.30
Morning stables	"	"	8 A.M.	"	9.0
Evening stables	"	"	4 P.M.	"	3.0
"Rack up"	"	"	8 P.M.		
Tattoo[1]	"	"	9 P.M.	"	8.0

If there was an order for a mounted parade the drum beat—

1st drum—"To horse." The men turned out, under the eye of

[1] In those days written Tap-to, meaning that no more liquor was to be drawn.

the quartermaster and fell in before the stable door in rank entire. Officers then inspected their troops; and each troop was told off in three divisions. 1760.

2nd drum—"Preparative." By the Adjutant's order.

3rd drum—"A flam." The centre division stood fast; the right division advanced, and the left division reined back, each two horses' lengths.

4th drum—"A flam." The front and rear divisions passaged to right and left and covered off, thus forming the troop in three ranks.

5th drum—"A march." The quartermasters led the troops to their proper position in squadron.

6th drum—"A flam." Officers rode to their posts (troop-leaders on the flank of their troops), facing their troops.

7th drum—"A flam." The officers halted, and turned about to their proper front.

Then the word was given—"Take care" (which meant "Attention"). "Draw your swords;" and the regiment was thus ready to receive the three squadron standards, which were escorted on to the ground and posted in the ranks, in the centre of the three squadrons.

Each squadron was then told off into half-squadrons, into three divisions, into half-ranks, into fours, and into files. As there are many people who do not know how to tell off a squadron by fours, it may be as well to mention how it was done. The men were not numbered off, but the officer went down each rank, beginning at the right-hand man, and said to the first, "You are the right-hand man of ranks by fours." Then going on to the fourth he said, "You are the left-hand man of ranks by fours," and so on. Telling off by files was a simpler affair. The officer rode down the ranks, pointing to each man, and saying alternately, "You move," "You stand," "You move," "You stand." Conceive what the confusion must have been if the men took it into their heads to be troublesome. "Beg your honour's pardon, but you said I was to stand," is the kind of speech that must have been heard pretty often in those days, when field movements went awry.

If the mounted parade went no further, the men marched back

1760. to their quarters in fours, each of the three ranks separately ; for in those days "fours" meant four men of one rank abreast. If field movements were practised, the system and execution thereof were left to the Colonel, unhampered by a drill-book. There was, however, a batch of "evolutions" which were prescribed by regulation, and required of every regiment when inspected by the King or a general officer. As these "evolutions" lasted, with some modification, till the end of the century, and (such is human nature) formed sometimes the only instruction, besides the manual exercise, that was imparted to the regiment, it may be as well to give a brief description thereof in this place. The efficiency of a regiment was judged mainly from its performance of the evolutions, which were supposed to be a searching test of horsemanship, drill, and discipline.

First then the squadron was drawn up in three ranks, at open order, that is to say, with a distance equal to half the front of the squadron between each rank. Then each rank was told off by half-rank, third of rank, and fours ; which done, the word was given, "Officers take your posts of exercise," which signified that the officers were to fall out to their front, and take post ten paces in rear of the commanding officer, facing towards the regiment. In other words, the regiment was required to go through the coming movements without troop or squadron leaders. Then the caution was given, "Take care to perform your evolutions," and the evolutions began.

To avoid tedium an abridgment of the whole performance is given at some length in the Appendix, and it is sufficient to say here that the first two evolutions consisted in the doubling of the depth of the column. The left half-ranks reined back and passaged to the right until they covered the right half-ranks ; and the original formation having been restored by more passaging, the right half-ranks did likewise. The next evolution was the conversion of three ranks into two, which was effected by the simple process of wheeling the rear rank into column of two ranks, and bringing it up to the flank of the front and centre ranks. Then came further variations of wheeling, and wheeling about by half-ranks, thirds of ranks, and fours; each movement being executed of course to the halt on a fixed pivot, so that through all these intricate manœuvres the regiment practically never moved off its ground. No doubt when performed, as in smart regiments they were performed, like clockwork, these evolutions were very pretty—and of course, like all drill, they had a disciplinary as well as an æsthetic value; but it must be confessed that they left a blight upon the British cavalry for more than a century. It is only within the last twenty years that the influence of these evolutions, themselves a survival from the days of Alexander the Great, has been wholly purged from our cavalry drill-books.

1760.

Meanwhile at this time (and for full forty years after for that matter) an immense deal of time was given up to dismounted drill; for the dragoons had not yet lost their character of mounted infantry. To dismount a squadron, the even numbers (as we should now say) reined back and passaged to the right; and the horses were then linked with "linking reins" carried for the purpose, and left in charge of the two flank men, while the rest on receiving the word, "Squadrons have a care to march forward," formed up in front, infantry wise, and were called for the time a battalion. This dismounted drill formed as important a feature of an inspection as the work done on horseback. Probably the survival of the march past the inspecting officer on foot may be traced to the traditions of those days.

If it be asked how time was found for so much dismounted work, the explanation is simple. From the 1st of May to the 1st October the troop horses were turned out to grass, and committed to the keeping of a "grass guard"—having, most probably, first gone through a course of bleeding at the hands of the farriers. It appears to have mattered but little how far distant the grass might be from the men's quarters; for we find that if it lay six or eight miles away, the "grass guard" was to consist of a corporal and six men, while if it were within a mile or two, two

1760 or three old soldiers were held to be amply sufficient. Men on "grass guard" were not allowed to take their cloaks with them, but were provided with special coats, whereof three or four were kept in each troop for the purpose. "Grass-time," it may be added, was not the busy, but the slack time for cavalrymen in those days—the one season wherein furloughs were permitted.

The close of the "grass-time" must have been a curious period in the soldier's year, with its renewal of the long-abandoned stable work and probable extra tightening of discipline. On the farriers above all it must have borne heavily, bringing with it, as we must conclude, the prospect of reshoeing every horse in the regiment. Moreover, the penalty paid by a farrier who lamed a horse was brutally simple: his liquor was stopped till the horse was sound. Nevertheless the farrier had his consolations, for he received a halfpenny a day for every horse under his charge, and must therefore have rejoiced to see his troop stable well filled. The men, probably, in a good regiment, required less smartening after grass-time than their horses. Light Dragoons thought a great deal of themselves, and were well looked after even on furlough. At the bottom of every furlough paper was a note requesting any officer who might read it to report to the regiment if the bearer were "unsoldierly in dress or manner." We gather, from a stray order, "that soldiers shall wear their hair *under* their hats," that even in those days men were bitten with the still prevailing fashion of making much of their hair; but we must hope that Hale's regiment knew better than to yield to it.

Every man, of course, had a queue of leather or of his own hair, either hanging at full length, in which case it was a "queue," or partly doubled back, when it became a "club." Which fashion was favoured by Colonel Hale we are, alas! unable to say,[1] but we gain some knowledge of the *coiffure* of the Light Dragoons from the following standing orders:—

[1] There were curious ideas afloat in those days about soldiers' heads. Colonel Dalrymple of the King's Own Dragoons suggests (1761) that the men's hair should be cut close, but that they should be provided with Spanish lamb's-wool wigs for cold and rainy weather.

"The Light Dragoon is always to appear clean and dressed in a soldier-like manner in the streets; his skirts tucked back, a black stock and black gaiters, but *no powder*. On Sundays the men are to have white stocks, and be well powdered, but no grease on their hair." 1760.

Here, therefore, we have a glimpse of the original trooper of the Seventeenth in his very best: his scarlet coat and white facings neat and spotless, the skirts tucked back to show the white lining, the glory of his white waistcoat, and the sheen of his white breeches. "Russia linen," *i.e.* white duck, would be probably the material of these last—Russia linen, "which lasts as long as leather and costs but half-a-crown," to quote one of our best authorities. Then below the white ducks, fitting close to the leg, came a neat pair of black cloth gaiters running down to dull black shoes, cleaned with "black ball" according to the regimental recipe. Round on his neck was a spotless white stock, helping, with the powder on his hair, to heighten the colour of his round, clean-shaven face. Very attractive he must have seemed to the girls of Coventry in the spring of 1760. What would we not give for his portrait by Hogarth as he appeared some fine Sunday in Coventry streets, with the lady of his choice on his arm, explaining to her that in the Light Dragoons they put no grease on their heads, and in proof thereof shaking a shower of powder from his hair on to her dainty white cap! Probably there were tender leave-takings when in September the regiment was ordered northward; possibly there are descendants of these men, not necessarily bearing their names, in Coventry to this day.

CHAPTER III

REFORMS AFTER THE PEACE OF PARIS, 1763-1774

1760. IN September Hale's Light Dragoons moved up to Berwick-on-Tweed, and thence into Scotland, where they were appointed to remain for the three ensuing years. Before it left Coventry the regiment, in common with all Light Dragoon regiments, had gathered fresh importance for itself from the magnificent behaviour of the 15th at Emsdorf on the 16th July; in which engagement Captain Martin Basil, who had returned to his own corps from Colonel Hale's, was among the slain. The close of the year brings us to the earliest of the regimental muster-rolls, which is dated Haddington, 8th December 1760. One must speak of muster-rolls in the plural, for there is a separate muster-roll for each troop—regimental rolls being at this period unknown.

These first rolls are somewhat of a curiosity, for that every one of them describes Hale's regiment as the 17th, the officers being evidently unwilling to yield seniority to the two paltry troops 1761. raised by Lord Aberdour. The next muster-rolls show considerable difference of opinion as to the regimental number, the head-quarter troop calling itself of the 18th, while the rest still claim 1762. to be of the 17th. In 1762 for the first time every troop 1763. acknowledges itself to be of the 18th, but in April 1763 the old conflict of opinion reappears; the head-quarter troop writes itself down as of the 18th, two other troops as of the 17th, while the remainder decline to commit themselves to any number at all. A gap in the rolls from 1763-1771 prevents us from following the controversy any further; but from this year 1763, the Seventeenth, as shall be shown, enjoys undisputed right to the number which it originally claimed. 1763.

Albeit raised for service in the Seven Years' War, the regiment was never sent abroad, though it furnished a draft of fifty men and horses to the army under Prince Ferdinand of Brunswick. All efforts to discover anything about this draft have proved fruitless; though from the circumstance that Lieutenant Wallop is described in the muster-rolls as "prisoner of war to the French," it is just possible that it served as an independent unit, and was actively engaged. But the war came to an end with the Treaty of Paris early in 1763; and with the peace came a variety of important changes for the Army, and particularly for the Light Dragoons.

The first change, of course, was a great reduction of the military establishment. Many regiments were disbanded—Lord Aberdour's, the 20th and 21st Light Dragoons among them. Colonel Hale's regiment was retained, and became the Seventeenth; and, as if to warrant it continued life, Hale himself was promoted to be full Colonel. We must not omit to mention here that, whether on account of his advancement, or from other simpler causes, Colonel Hale in this same year took to himself a wife, Miss Mary Chaloner of Guisbrough. History does not relate whether the occasion was duly celebrated by the regiment, either at the Colonel's expense or at its own; but it is safe to assume that, in those hard-drinking days, such an opportunity for extra consumption of liquor was not neglected. If the fulness of the quiver be accepted as the measure of wedded happiness, then we may fearlessly assert that Colonel Hale was a happy man. Mrs. Hale bore him no fewer than twenty-one children, seventeen of whom survived him.

The actual command of the regiment upon Colonel Hale's promotion devolved upon Lieut.-Colonel Blaquiere, whose duty it now became to carry out a number of new regulations laid down after the peace for the guidance of the Light Dragoons. By July 1764 these reforms were finally completed; and as they remained 1764

1764. in force for another twenty years, they must be given here at some length. The pith of them lies in the fact that the authorities had determined to emphasise in every possible way the distinction between Light and Heavy Cavalry. Let us begin with the least important, but most sentimental of all matters—the dress.

PRIVATES

Coat.—(Alike for all ranks.) Scarlet, with 3-inch white lapels to the waist. White collar and cuffs, sleeves unslit. White lining. Braid on button-holes. Buttons, in pairs, white metal with regimental number.

Waistcoat.—White, unembroidered and unlaced. Cross pockets.

Breeches.—White, duck or leather.

Boots.—To the knee, "round toed and of a light sort."

Helmet.—Black leather, with badge of white metal in front, and white turban round the base, plume and crest scarlet and white.

Forage Cap.—Red, turned up with white. Regimental number on little flap.

Shoulder Belts.—White, $2\frac{3}{4}$ inches broad. Sword belt over the right shoulder.

Waist Belt.—White, $1\frac{3}{4}$ inches broad.

Cloaks.—Red, white lining; loop of black and white lace on the top. White cape.

Epaulettes.—White cloth with white worsted fringe.

CORPORALS

Same as the men. Distinguished by narrow silver lace round the turn-up of the sleeves. Epaulettes bound with white silk tape, white silk fringe.

SERGEANTS

Same as the men. Epaulettes bound with narrow silver lace; silver fringe. Narrow silver lace round button-holes. Sash of spun silk, crimson with white stripe.

QUARTERMASTERS

Same as the men. Silver epaulettes. Sash of spun silk, crimson.

OFFICERS

Same as the men; but with silver lace or embroidery at the Colonel's discretion. Silk sash, crimson. Silver epaulettes. Scarlet velvet stock and waist belts. 1764.

Trumpeters

White coats with scarlet lapels and lining; lace, white with black edge; red waistcoats and breeches. Hats, cocked, with white plume.

Farriers

Blue coats, waistcoats, and breeches. Linings and lapels blue; turn-up of sleeves white. Hat, small black bearskin, with a horse-shoe of silver-plated metal on a black ground. White apron rolled back on left side.

Horse Furniture.—White cloth holster caps and housings bordered with white, black-edged lace. XVII. L. D. embroidered on the housings on a scarlet ground, within a wreath of roses and thistles. King's cypher, with crown over it and XVII. L. D. under it embroidered on the holster caps.

Officers had a silver tassel on the holster caps and at the corners of the housings.

Quartermasters had the same furniture as the officers, but with narrower lace, and without tassels to the holster caps.

Arms

Officers.—A pair of pistols with barrels 9 inches long. Sword (straight or curved according to regimental pattern), blade 36 inches long. A smaller sword, with 28-inch blade, worn in a waist belt, for foot duty.

Men.—Sword and pistols, as the officers. Carbine, 2 feet 5 inches long in the barrel. Bayonet, 12 inches long. Carbine and pistols of the same bore. Cartridge-box to hold twenty-four rounds.

So much for the outward adornment and armament of the men, to which we have only to add that trumpeters, to give them further distinction, were mounted on white horses, and carried a sword with a scimitar blade. Farriers, who were a peculiar people in those days, were made as dusky as the trumpeters were gorgeous. They carried two churns instead of holsters on their saddles, wherein to stow their shoeing tools, etc., and black bearskin furniture with crossed hammer and pincers on the housing. Their weapon was an axe, carried, like the men's swords, in a belt

1764. slung from the right shoulder. When the men drew swords, the farriers drew axes and carried them at the "advance." The old traditions of the original farrier still survive in the blue tunics, black plumes, and axes of the farriers of the Life Guards, as well as in the blue stable jackets of their brethren of the Dragoons.

Passing now from man to horse, we must note that from 27th July 1764 it was ordained that the horses of Horse and Dragoons should in future wear their full tails, and that those of Light Dragoons only should be docked.[1] This was the first step towards the reduction of the weight to be carried by the Light Dragoon horse. The next was more practical. A saddle much lighter than the old pattern was invented, approved, and adopted, with excellent results. It was of rather peculiar construction : very high in the pommel and cantle, and very deep sunk in the seat, in order to give a man a steadier seat when firing from on horseback. Behind the saddle was a flat board or tray, on to which the kit was strapped in a rather bulky bundle. It was reckoned that this saddle, with blanket and kit complete, 30 lbs of hay and 5 pecks of oats, weighed just over 10 stone (141 lbs.); and that the Dragoon with three days' rations, ammunition, etc., weighed 12 stone 7 lbs. more ; and that thus the total weight of a Dragoon in heavy marching order with (roughly speaking) three days' rations for man and horse, was 22 stone 8 lbs. In marching from quarter to quarter in England, the utmost weight on a horse's back was reckoned not to exceed 16 stone.

A few odd points remain to be noticed before the question of saddlery is finally dismissed. In the first place, there was rather an uncouth mixture of colours in the leather, which, though designed to look well with the horse furniture, cannot have been beautiful without it. Thus the head collar for ordinary occasions was brown, but for reviews white ; bridoons were black, bits of bright steel ; the saddle was brown, and the carbine bucket black. These buckets were, of course, little more than leather caps five

[1] They were said, when thus docked, to have "hunter's tails"; hence, perhaps, the popular identification of the Light Dragoon officer with the sportsman.

or six inches long, fitting over the muzzle of the carbine, practically the same as were served out to Her Majesty's Auxiliary Cavalry less than twenty years ago. Light Dragoons, however, had a swivel fitted to their shoulder-belt to which the carbine could be sprung, and the weapon thus made more readily available. The horse furniture of the men was not designed for ornament only; for, being made in one piece, it served to cover the men when encamped under canvas. As a last minute point, let it be noted that the stirrups of the officers were square, and of the men round at the top.

1764.

We must take notice next of a more significant reform, namely, the abolition of side drums and drummers in the Light Dragoons, and the substitution of trumpeters in their place. By this change the Light Dragoons gained an accession of dignity, and took equal rank with the horse of old days. The establishment of trumpeters was, of course, one to each troop, making six in all. When dismounted they formed a "band of music," consisting of two French horns, two clarionets, and two bassoons, which, considering the difficulties and imperfections of those instruments as they existed a century and a quarter ago, must have produced some rather remarkable combinations of sound. None the less we have here the germ of the regimental band, which now enjoys so high a reputation.

Over and above the trumpeters, the regiment enjoyed the possession of a fife, to whose music the men used to march. At inspection the trumpets used to sound while the inspecting officer went down the line; and when the trumpeters could blow no longer, the fife took up the wondrous tale and filled up the interval with an ear-piercing solo. The old trumpet "marches" are still heard (unless I am mistaken) when the Household Cavalry relieve guard at Whitehall. But more important than these parade trumpet sounds is the increased use of the trumpet for signalling movements in the field. The original number of trumpet-calls in the earliest days of the British cavalry was, as has already been mentioned, but six. These six were apparently

1764. still retained and made to serve for more purposes than one; but others also were added to them. And since, so far as we can gather, the variety of calls on one instrument that could be played and remembered was limited by human unskilfulness and human stupidity, this difficulty was overcome by the employment of other instruments. These last were the bugle horn and the French horn; the former the simple curved horn that is still portrayed on the appointments of Light Infantry, the latter the curved French hunting horn. The united efforts of trumpet, bugle horn, and French horn availed to produce the following sounds:—

	Stable call—Trumpet.
(*Butte Sella*).[1]	Boot and saddle—Trumpet.
(*Monte Cavallo*).[1]	Horse and away—Trumpet. But sometimes bugle horn; used also for evening stables.
(? *Tucquet*).[1]	March—Trumpet.
	Water—Trumpet.
(*Auquet*).[1]	Setting watch or tattoo—Trumpet. Used also for morning stables.
(? *Tucquet*).[1]	The call—Trumpet. Used for parade or assembly.
	Repair to alarm post—Bugle horn.
(*Alla Standarda*).[1]	Standard call—Trumpet. Used for fetching and lodging standards; and also for drawing and returning swords.
	Preparative for firing—Trumpet.
	Cease firing—Trumpet.
	Form squadrons, form the line—Bugle horn.
	Advance—Trumpet.
(*Carga*).[1]	Charge or attack—Trumpet.
	Retreat—French horns.
	Trot, gallop, front form—Trumpet.
	Rally—Bugle horn.
	Non-commissioned officers' call—Trumpet.

The quick march on foot—The fife.
The slow march on foot—The band of music.

All attempts to discover the notation of these calls have, I regret to say, proved fruitless, so that I am unable to state

[1] Denotes one of the six original trumpet-calls.

positively whether any of them continue in use at the present day. The earliest musical notation of the trumpet sounds that I have been able to discover dates from the beginning of this century,[1] and is practically the same as that in the cavalry drill-book of 1894; so that it is not unreasonable to infer that the sounds have been little altered since their first introduction. Indeed, it seems to me highly probable that the old "Alla Standarda," which is easily traceable back to the first quarter of the seventeenth century, still survives in the flourish now played after the general salute to an inspecting officer. As to the actual employment of the three signalling instruments in the field, we shall be able to judge better while treating of the next reform of 1763-1764, viz. that of the drill.

1764.

The first great change wrought by the experience of the Seven Years' War on the English Light Dragoon drill was the final abolition of the formation in three ranks. Henceforward we shall never find the Seventeenth ranked more than two deep. Further, we find a general tendency to less stiffness and greater flexibility of movement, and to greater rapidity of manœuvre. The very evolutions sacrifice some of their prettiness and precision in order to gain swifter change of formation. Thus, when the left half rank is doubled in rear of the right, the right, instead of standing fast, advances and inclines to the left, while the latter reins back and passages to the right, thus accomplishing the desired result in half the time. Field manœuvres are carried out chiefly by means of small flexible columns, differing from the present in one principal feature only, viz. that the rear rank in 1763 does not inseparably follow the front rank, but that each rank wheels from line into column of half-ranks or quarter-ranks independently. Moreover, we find one great principle pervading all field movements: that Light Dragoons, for the dignity of their name, must move with uncommon rapidity and smartness. The very word "smart," as applied to the action of a soldier, appears, so far as I know, for the first time in a drill-book made for Light

[1] The calls were first authorised by regulation (so far as is known) in 1799.

1764. Dragoons at this period. In illustration, let us briefly describe a parade attack movement, which is particularly characteristic.

The regiment having been formed by previous manœuvres in echelon of wings (three troops to a wing) from the left, the word is given, "Advance and gain the flank of the enemy."

First Trumpet.—The right files (of troops?) of each wing gallop to the front, and form rank entire; unswivel their carbines, and keep up a rapid irregular fire from the saddle.

Under cover of this fire the echelon advances.

Second Trumpet.—The right wing forms the "half-wedge" (single echelon), passes the left or leading wing at an increased pace, and gains the flank of the imaginary enemy by the "head to haunch" (an extremely oblique form of incline), and forms line on the flank.

Third Trumpet—"*Charge.*"—The skirmishers gallop back through the intervals to the rear of their own troops, and remain there till the charge is over.

French Horns—"*Retreat.*"—The skirmishers gallop forward once more, and keep up their fire till the line is reformed.

The whole scheme of this attack is perhaps a shade theatrical, and, indeed, may possibly have been designed to astonish the weak mind of some gouty old infantry general; but a regiment that could execute it smartly could hardly have been in a very inefficient state.

1765. In 1765 the Seventeenth was moved to Ireland, though to what part of Ireland the gap in the muster-rolls disenables us to say. Almost certainly it was split up into detachments, where we have reason to believe that the troop officers took pains to teach their men the new drill. We must conceive of the regiment's life as best we may during this period, for we have no information to help us. Colonel Blaquiere, we have no doubt, paid visits to the outlying troops from time to time, and probably was able now and again to get them together for work in the field, particularly when an inspecting officer's visit was at hand. We know, from the inspection returns, that the Seventeenth advanced and gained the flank of the enemy every year, in a fashion which commanded the admiration of all beholders. And let us note that in this very year the British Parliament passed an Act for the imposition of stamp duties on the American Colonies—preparing, though unconsciously, future work on active service for the Seventeenth. 1765.

For the three ensuing years we find little that is worth the chronicling, except that in 1766 the regiment suffered, for a brief period, a further change in its nomenclature, the 15th, 16th, and 17th being renumbered the 1st, 2nd, and 3rd Light Dragoons. In this same year we discover, quite by chance, that two troops of the Seventeenth were quartered in the Isle of Man, for how long we know not. In 1767 a small matter crops up which throws a curious light on the grievances of the soldier in those days. Bread was so dear that Government was compelled to help the men to pay for it, and to ordain that on payment of fivepence every man should receive a six-pound loaf—which loaf was to last him for four days. Let us note also, as a matter of interest to Colonel Blaquiere, a rise in the value of another article, namely, the troop horse, whereof the outside price was in this year raised from twenty to twenty-two guineas. 1766.

In 1770 we find Colonel Hale promoted to be Governor of Limerick, and therewith severed from the regiment which he had raised. As his new post must presumably have brought him over to Ireland, we may guess that the regiment may have had an opportunity of giving him a farewell dinner, and, as was the fashion in those days, of getting more than ordinarily drunk. From this time forward we lose sight of Colonel Hale, though he is still a young and vigorous man, and has thirty-three years of life before him. His very name perishes from the regiment, for if ever he had an idea of placing a son therein, that hope must have been killed long before the arrival of his twenty-first child. His successor in the colonelcy was Colonel George Preston of the Scots Greys, a distinguished officer who had served at Dettingen, Fontenoy, and other actions of the war of 1743-47, as well as in the principal battles of the Seven Years' War. 1770.

Meanwhile, through all these years, the plot of the American

1770. dispute was thickening fast. From 1773 onwards the news of trouble and discontent across the Atlantic became more frequent; and at last in 1774 seven infantry regiments were despatched to Boston. Then probably the Seventeenth pricked up its ears and discussed, with the lightest of hearts, the prospect of fighting the 1775. rebels over the water. The year 1775 had hardly come in when the order arrived for the regiment to complete its establishment with drafts from the 12th and 18th, and hold itself in readiness to embark at Cork for the port of Boston. It was the first cavalry regiment selected for the service—a pretty good proof of its reputation for efficiency.[1]

[1] These are fragments of some of the inspection reports:—1770, "A *very good* regiment." 1771, "A very fine regiment, and appears perfectly fit for service. Must have had great care taken of it." 1772, "In every respect a fine regiment and fit for service." 1773, "This regiment is an extreme pretty one and in good order." 1774, "This regiment is in great order and fit for service."

CHAPTER IV

THE AMERICAN WAR—1ST STAGE—THE NORTHERN CAMPAIGN, 1775-1780.

It would be beside the purpose to enter upon a relation of the causes which led to the rupture between England and the thirteen North American Colonies, and to the war of American Independence. The immediate ground of dispute was, however, one in which the Army was specially interested, namely, the question of Imperial defence. Fifteen years before the outbreak of the American War England had, by the conquest of Canada, relieved the Colonies from the presence of a dangerous neighbour on their northern frontier, and for this good service she felt justified in asking from them some return. Unfortunately, however, the British Government, instead of leaving it to the Colonies to determine in what manner their contribution to the cost of Imperial defence should be raised, took the settlement of the question into its own hands, as a matter wherein its authority was paramount. Ultimately by a series of lamentable blunders the British ministers contrived to create such irritation in America that the Colonies broke into open revolt. 1775.

It was in the year 1774 that American discontent reached its acutest stage; and the centre of that discontent was the city of Boston. In July General Gage, at that time in command of the forces in America, and later on to be Colonel-in-Chief of the 17th Light Dragoons, feeling that the security of Boston was now seriously threatened by the rebellious attitude of the citizens, moved down with some troops and occupied the neck of the 1774.

1774. isthmus on which the city stands. This step increased the irritation of the people so far that in a month or two he judged it prudent to entrench his position and remove all military stores from outlying stations into Boston. By November the temper of the Colonists had become so unmistakably insubordinate that Gage issued a proclamation warning them against the consequences of revolt. This manifesto was taken in effect as a final signal for general and open insurrection. Rhode Island and New Hampshire broke out at once; and the Americans began their military preparations by seizing British guns, stores, and ammunition

1775. wherever they could get hold of them. By the opening of 1775 the seizure, purchase, and collection of arms became so general that Gage took alarm for the safety of a large magazine at Concord, some twenty miles from Boston, and detached a force to secure it. This expedition it was that led to the first shedding of blood. The British troops succeeded in reaching Concord and destroying the stores; but they had to fight their way back to Boston through the whole population of the district, and finally arrived, worn out with fatigue, having lost 240 men, killed,

19th April. wounded, and missing, out of 1800. The Americans then suddenly assembled a force of 20,000 men and closely invested Boston.

It was just about this time that there arrived in Boston Captain Oliver Delancey, of the 17th Light Dragoons, with despatches announcing that reinforcements would shortly arrive from England under the command of Generals Howe and Clinton. Captain Delancey was charged with the duty of preparing for the reception of his regiment, and in particular of purchasing horses whereon to mount it. Two days after his arrival, therefore, he started for New York to buy horses, only to find at his journey's end that New York also had risen in insurrection, and that there was nothing for it but to return to Boston.

And while Delancey was making his arrangements, the Seventeenth was on its way to join him. The 12th and 18th Regiments had furnished the drafts required of them, and the Seventeenth, thus raised to some semblance of war strength, embarked for its first turn on active service. Here is a digest of their final muster, dated, Passage, 10th April 1775, and endorsed "Embarkation"—

1775.

10th April.

Lieutenant-Colonel.—Samuel Birch.

Major.—Henry Bishop.

Adjutant.—John St. Clair, *Cornet.*

Surgeon.—Christopher Johnston.

Surgeon's mate.—Alexander Acheson.

Deputy-Chaplain.—W. Oliver.

Major Bishopp's Troop.

Robert Archdale, *Captain.* Frederick Metzer, *Cornet.*
1 Quartermaster, 2 sergeants, 2 corporals, 1 trumpeter, 29 dragoons, 31 horses.

Captain Straubenzee's Troop.

Henry Nettles, *Lieutenant.* Sam. Baggot, *Cornet.*
5 Non-commissioned officers, 1 trumpeter, 26 dragoons, 31 horses.

Captain Moxham's Troop.

Ben. Bunbury, *Lieutenant.* Thomas Cooke, *Cornet.*
5 Non-commissioned officers, 1 trumpeter, 26 dragoons, 31 horses.

Captain Delancey's Troop.

Hamlet Obins, *Lieutenant.* James Hussey, *Cornet.*
5 Non-commissioned officers, 1 trumpeter, 1 hautboy, 27 dragoons, 31 horses.

Captain Needham's Troop.

Mark Kerr, *Lieutenant.* Will. Loftus, *Cornet.*
5 Non-commissioned officers, 1 trumpeter, 26 dragoons, 31 horses.

Captain Crewe's Troop.

Matthew Patteshall, *Lieutenant.* John St. Clair (Adjutant), *Cornet.*
5 Non-commissioned officers, 1 trumpeter, 1 hautboy, 26 dragoons, 31 horses.

1775. What manner of scenes there may have been at the embarkation that day at Cork it is impossible to conjecture. We can only bear in mind that there were a great many Irishmen in the ranks, and that probably all their relations came to see them off, and draw what mental picture we may. Meanwhile it is worth while to compare two embarkations of the regiment on active service, at roughly speaking, a century's interval. In 1879 the Seventeenth with its horses sailed to the Cape in two hired transports—the *England* and the *France*. In 1776 it filled no fewer than seven ships, the *Glen*, *Satisfaction*, *John and Jane*, *Charming Polly*, *John and Rebecca*, *Love and Charity*, *Henry and Edward*—whereof the very names suffice to show that they were decidedly small craft.

The voyage across the Atlantic occupied two whole months, but, like all things, it came to an end; and the regiment dis-

June 15-19. embarked at Boston just in time to volunteer its services for the first serious action of the war. That action was brought about in this way. Over against Boston, and divided from it by a river of about the breadth of the Thames at London Bridge, is a peninsula called Charlestown. It occurred, rather late in the day, to General Gage that an eminence thereupon called Bunker's Hill was a position that ought to be occupied, inasmuch as it lay within cannon-shot of Boston and commanded the whole of the town. Unfortunately, precisely the same idea had occurred to the Americans, who on the 16th June seized the hill, unobserved by Gage, and proceeded to entrench it. By hard work and the aid of professional engineers they soon made Bunker's Hill into a formidable position; so that Gage, on the following day, found that his task was not that of marching to an unoccupied height, but of attacking an enemy 6000 strong in a well-fortified post. None the less he attacked the 6000 Americans with 2000 English, and drove them out at the bayonet's point after the bloodiest engagement thitherto fought by the British army. Of the 2000 men 1054, including 89 officers, went down that day; and the British occupied the Charlestown peninsula.

The acquisition was welcome, for the army was sadly crowded in Boston and needed more space; but the enemy soon erected new works which penned it up as closely as ever. Moreover the Americans refused to supply the British with fresh provisions, so that the latter—what with salt food, confinement, and the heat of the climate—soon became sickly. The Seventeenth were driven to their wit's end to obtain forage for their horses. It was but a poor exchange alike for animals and men to forsake the ships for a besieged city. The summer passed away and the winter came on. The Americans pressed the British garrison more hardly than ever through the winter months, and finally, on the 2nd March 1776, opened a bombardment which fairly drove the English out. On the 17th March Boston was evacuated, and the army, 9000 strong, withdrawn by sea to Halifax.

1775.

1776.

However mortifying it might be to British sentiment, this evacuation was decidedly a wise and prudent step; indeed, but for the determination of King George III. to punish the recalcitrant Boston, it is probable that it would have taken place long before, for it was recommended both by Gage, who resigned his command in August 1775, and by his successor, General Howe. They both saw clearly enough that, as England held command of the sea, her true policy was to occupy the line of the Hudson River from New York in the south to Lake Champlain in the north. Thereby she could isolate from the rest the seven provinces of Connecticut, Rhode Island, Massachusetts, Vermont, New Hampshire, and Maine, and reduce them at her leisure; which process would be the easier, inasmuch as these provinces depended almost entirely on the States west of the Hudson for their supplies. The Americans, being equally well aware of this, and having already possession of New York, took the bold line of attempting to capture Canada while the English were frittering their strength away at Boston. And they were within an ace of success. As early as May 1775 they captured Ticonderoga and the only King's ship in Lake Champlain, and in November they obtained possession of

1776. Chambly, St. John's, and Montreal. Fortunately Quebec still held out, though reduced to great straits, and saved Canada to England. On the 31st December the little garrison gallantly repelled an American assault, and shortly after it was relieved by the arrival of a British squadron which made its way through the ice with reinforcements of 3500 men under General Burgoyne. This decided the fate of Canada, from which the Americans were finally driven out in June 1776.

One other small incident requires notice before we pass to the operations of Howe's army (whereof the Seventeenth formed part) in the campaign of 1776. Very early in the day Governor Martin of North Carolina had recommended the despatch of a flying column or small force to the Carolinas, there to rally around it the loyalists, who were said to be many, and create a powerful diversion in England's favour. Accordingly in December 1775, five infantry regiments under Lord Cornwallis were despatched from England to Cape Fear, whither General Clinton was sent by Howe to meet them and take command. An attack on Charleston by this expedition proved to be a total failure; and on the 21st June 1776, Clinton withdrew the force to New York. This episode deserves mention, because it shows how early the British Government was bitten with this plan of a Carolina campaign, which was destined to cost us the possession of the American Colonies. Three times in the course of this history shall we see English statesmen make the fatal mistake of sending a weak force to a hostile country in reliance on the support of a section of disaffected inhabitants, and each time (as fate ordained it) we shall find the Seventeenth among the regiments that paid the inevitable penalty. From this brief digression let us now return to the army under General Howe.

While the bulk of this force was quartered at Halifax, the Seventeenth lay, for convenience of obtaining forage, at Windsor, some miles away. In June the 16th Light Dragoons arrived at Halifax from England with remounts for the regiment; but it is questionable whether they had any horses to spare, for we find that out of 950 horses 412 perished on the voyage. 1776. About the same time arrived orders for the increase of the Seventeenth by 1 cornet, 1 sergeant, 2 corporals, and 30 privates per troop; but the necessary recruits had not been received by the time when the campaign opened. On the 11th June the regiment, with the rest of Howe's army, was once more embarked at Halifax and reached Sandy Hook on the 29th. Howe then landed his force on Staten Island, and awaited the arrival of his brother, Admiral Lord Howe, who duly appeared with a squadron and reinforcements on the 1st July. Clinton with his troops from Charleston arrived on the 1st August, and further reinforcements from England on the 12th. Howe had now 30,000 men, 12,000 of them Hessians, under his command in America, two-thirds of whom were actually on the spot around New York.

Active operations were opened on the 22nd August, by the landing of the whole army in Gravesend Bay at the extreme south-west corner of Long Island. The American army, 15,000 strong, occupied a position on the peninsula to the north-west, where Brooklyn now stands—its left resting on the East River, its right on a stream called Mill Creek, and its front covered as usual by a strong line of entrenchments. From this fortified camp, however, they detached General Putnam with 10,000 men to take up a position about a mile distant on a line of heights that runs obliquely across the island. After a reconnaissance by Generals Clinton and Erskine, the latter of whom led the brigade to which the Seventeenth was attached, General Howe decided to turn the left flank of the Americans with part of his force, leaving the rest to attack their front as soon as the turning movement was completed. At 9 P.M. on the 26th August the turning column, under the command of Howe himself, marched across the flat ground to seize a pass on the extreme left of the enemy's line, the Seventeenth forming the advanced guard. On reaching the pass it was found that the Americans had neglected to secure it, being content to visit it with occasional cavalry

1776. patrols. One such patrol was intercepted by the advanced party of the Seventeenth; and the pass was occupied by the British without giving alarm to the Americans. At nine next morning, Howe's column having completely enveloped Putnam's left, opened the attack on that quarter, while the rest of the army advanced upon the centre and right. The Americans were defeated at all points and driven in confusion to their entrenchments; but Howe made no effort to pursue them nor to storm the camp, as he might easily have done. He merely moved feebly up to the enemy's entrenchments on the following day, and began to break ground as if for a regular siege. On the 29th the Americans evacuated the camp, and retired across the East River to New York; and this they were allowed to do without hindrance, though the British army of 20,000 men stood on their front, and a navigable river, where a British seventy-four could have anchored, lay in their rear. Thus deliberately were sacrificed the fruits of the battle of Brooklyn. This was the first action in which the Seventeenth was under fire. The regiment at its close received the thanks of Generals Erskine and Clinton.

The possession of Long Island gave the British complete command of New York by sea; and Howe set himself to transport his army to New York Island, an operation which was completed on the 15th September. The Americans then evacuated New York town and retired to the northern extremity of New York Island, where Washington fortified a position from Haarlem to Kingsbridge along the Hudson River in order to secure his retreat across it to the mainland. The English warships now moved up the Hudson to cut off that retreat; and Howe having left four brigades to cover New York town, 12th Oct. embarked the rest on flat-bottomed boats to turn Washington's position. The flotilla passed through Hell Gate; and Howe 18th Oct. having wasted a deal of time in disembarking the troops first at the wrong place, landed them finally at Pell's Point, the corner which divides East River from Long Island Sound, and forms the extreme point of the spit of continent that runs down to New York Island. The advanced parties of the Seventeenth were engaged in a trifling skirmish at Pelham Manor, a little to the north of Pell's Point, shortly after disembarkation; but the British advance was practically unopposed, and the army was concentrated at New Rochelle, on Long Island Sound, on the 21st October. Washington now changed front, throwing his left back, and distributed his army along a line parallel to the march of the British; his right resting at Kingsbridge on the south, and his left at Whiteplains on the north. The two armies were separated by a deep river called the Bronx, which covered the whole of Washington's front. Howe continued his march northward, doubtless with the intention of getting between Washington and the mainland; but Washington had already sent parties to entrench a new position for him at Whiteplains, to which he moved on the 26th October. This change of position brought the Americans from the left flank to the front of the British advance, and it was plain that an action was imminent. On the 28th, Howe's army, advancing in two columns, came up with the Americans, and found them to be some 18,000 strong. The right of Washington's main position rested on the Bronx River; but for some reason a detached force of 4000 men had been posted on a hill on the other side of the river, which detachment, owing to the depth and difficulty of the stream, was necessarily cut off from the rest of the line. Howe decided to attack this isolated body at once. The Seventeenth being detailed as part of the attacking force, moved off to a practicable ford, the passage of which was carried in the face of heavy fire; and the infantry then advancing drove the enemy brilliantly from their entrenchments, from whence the Seventeenth pursued them towards the main position at Whiteplains. The regiment lost one man and five horses killed, Cornet Loftus, four men and eight horses wounded, in this action; which unfortunately led to no result. On the 30th August a general attack on the American entrenchments was ordered, but

1776.

1776. was countermanded in consequence of a tremendous storm of rain; and on the 1st September the Americans quietly retired northward across the river Croton, on which they took up a position from which it was hopeless to attempt to dislodge them.

However, there was still an American garrison of 3000 men, which had been left by Washington in his entrenchments at Kingsbridge to hold the passage of the Hudson; and of these Howe determined to make sure. His attack was delivered by four columns simultaneously. The third of these crossed the Haarlem Creek in boats under a heavy fire, and by the capture of a strong post at the other side turned the left of the American position. The ground was unfavourable for cavalry, however; and the Seventeenth, which was attached to this column, lost but one man. The result of the whole operation was the surrender of the Americans, which was bought with the loss of 800 British killed and wounded.

Three days later Lord Cornwallis crossed the Hudson with 4000 men, and marched against the American fort which commanded the passage of the river from the Jersey side. The Americans promptly evacuated it and retreated, with Cornwallis at their heels in hot pursuit. He was on the point of overtaking them and striking a severe blow, when he received orders from General Howe to halt—orders which he very reluctantly obeyed. A party of the Seventeenth, probably a sergeant's party for orderly duties, seems to have accompanied Cornwallis on this march, and through the gallant behaviour of one of the men has made itself remembered.

One day Private M'Mullins, of this detachment, was despatched by Lord Cornwallis with a letter of some importance to an officer of one of the outposts, and while passing near a thicket on his way was fired at by the rebels. He instantly pretended to fall from his horse, hanging with head down to the ground. The Americans, four in number, supposing him killed, ran out from their cover to seize their booty, and had come within a few yards of him, when, to their great astonishment, Private M'Mullins suddenly recovered his seat in the saddle and shot the first of them dead with his carbine. He then drew his pistol and despatched a second, and immediately after fell with his sword upon the other two, who surrendered as his prisoners. Whereupon Private M'Mullins drove them triumphantly before him into camp, where he duly delivered them up. Lord Cornwallis did not fail to report such bravery to General Howe, who in his turn not only promoted M'Mullins to be sergeant, but brought the exploit before the notice of the King. As all Light Dragoons of whatever regiment felt pride in their comrades, the story of Private M'Mullins found its way into the standard contemporary work on that branch of the service, and remains there embalmed to this day. Let it be noted that this feat of leaning out of the saddle almost to the ground is treated as one which "all Light Dragoons accomplished with the greatest ease." We should probably never have known this but for Private M'Mullins of the Seventeenth.

1776.

With the recall of Cornwallis from New Jersey the campaign of 1776 came to an end. Since the American evacuation of New York, Howe had captured 4500 prisoners and 150 guns; but he had also thrice let slip the opportunity of capturing the whole American army. One further operation was insisted upon by the Admiral, namely, the capture of Rhode Island, which was effected without loss by a small force under General Clinton. One troop of the Seventeenth accompanied Clinton on this expedition, and remained at Rhode Island for the next twelve months.

8th Dec.

The rest of the Seventeenth went into winter quarters in New York, the total strength of the regiment at the close of the campaign being 225 men. Though its casualties had been light, it had done a good deal of hard work and established for itself a reputation. Howe himself testifies in his despatches to "the good service they have performed in this campaign," and adds that "the dread which the enemy have of the Dragoons has been experienced on every occasion." It is a significant indication of

1777. the nature of their work, that Howe begs for remounts of Irish horses for them, as being "hardier and better accustomed to get over fences."

The rest of the army in the winter of 1776-77 was split up into detachments, and scattered along an extended line from the Delaware to New York. The Americans fully expected Howe to cross the Delaware as soon as the ice permitted and attack Philadelphia, but Howe as usual did nothing. He might have destroyed the American army without difficulty; but so far from attempting it, he allowed Washington with an inferior force to cut off two detached posts and do a great deal of damage.

Howe's operations in the campaign of 1777 were little more satisfactory. After making every preparation to cross the Delaware and advance into Pennsylvania he brought back the army to New York, and embarked for the Chesapeake in order to approach Philadelphia from that side. In September he won the battle of Brandywine, and took possession of Philadelphia on the 26th. This occupation of Philadelphia was the sole result of the campaign; and it was, in fact, a political rather than a military enterprise, the object being to overawe the American Congress. It was a fatal mistake, for while Howe was wasting his time in Pennsylvania, Burgoyne was moving down from Canada to open the line of the Hudson from the north, in the hope of co-operation from Howe's army in the south. No such co-operation was forthcoming. Howe's army was engaged elsewhere; Clinton, though, as will be seen, he did make on his own responsibility a slight diversion on the Hudson, yet dared not weaken the garrison of New York. The result was that 16th Oct. Burgoyne with his whole force of 7000 men was overpowered and compelled to surrender at Saratoga.

The Seventeenth being left in garrison at New York, of course took no share in Howe's operations. The fact was that in November 1776 it received some 200 recruits and 100 fresh horses from England, so that its time must have been fully occupied in the task of knocking these into shape. Nevertheless small detachments of the regiment were employed in two little affairs which must be related here. 1777.

The Americans, after retreating across the Croton in 1776, had formed large magazines on the borders of Connecticut, at the town of Danbury and elsewhere. These magazines General Clinton judged that it would be well to destroy. Accordingly, on the 25th April, 2000 men, drafted from different regiments, including twelve from the Seventeenth for the needful reconnaissance and patrol duties, embarked on transports and sailed up Long Island Sound to Camp's Point, where they landed. At ten that night they marched, and at eight next morning they reached Danbury, to the great surprise of the Americans, who evacuated the town with all speed. The British, having destroyed the whole of the stores, prepared to return to their ships, but found that the Americans had assembled at a place called Ridgefield, and had there entrenched themselves to bar the British line of march. Weary as they were after twenty-four hours' work, the English soldiers attacked and carried the entrenchments; and then, as night came on, they lay on their arms, prepared to fight at any moment. At daybreak they continued their march, and were again attacked by the Americans, who had received reinforcements during the night. Still they fought their way on till within half a mile of their ships, when General Erskine, losing all patience, collected 400 men, and taking the offensive at last beat the enemy off. The men had had no rest for three days and three nights, and were fairly worn out; but we may guess that the little detachment of the Seventeenth was not the last to answer to the call of its Brigadier. This expedition cost the British 15 officers and 153 men.

The second of the two affairs to which we have alluded was an expedition made by Clinton as a diversion to help Burgoyne, and was directed against two American forts on the right bank of the Hudson, which barred the passage of the British war-ships to Albany; Albany being the point to which Burgoyne hoped to penetrate. A force of 3000 men, including one troop of the

1777. 5th Oct. Seventeenth, embarked on the 5th October and sailed up the Hudson to Verplanks Point, forty miles from New York, on the east bank of the river. Here Clinton landed a portion of his force under the fire of a small American field-work, drove out the enemy, and pursued them for some little way. This feint produced the desired effect. The American general of the district at once concluded that Clinton meant to advance to meet Burgoyne on the east bank of the Hudson, and hurried away with most of the garrison of the river ports to occupy the passes on the roads. Clinton meanwhile quietly embarked two-thirds of his force on the following morning, leaving the remainder to hold Verplanks, and landed them on the opposite bank. Thence he advanced over a very steep mountain, along very bad roads, to attack two important posts, Forts Clinton and Montgomery, from the rear. Though Fort Clinton, the lower of the two, was but twelve miles distant, it was not reached before sunset, owing to the difficulties of the march. Opposite Fort Clinton the force divided into two columns, one of them standing fast, while the other made a detour to reach Fort Montgomery unobserved—the design being to attack both posts, which were only three-quarters of a mile apart, simultaneously. The upper post, Fort Montgomery, was easily captured, being at once abandoned by its garrison of 800 men. Fort Clinton, however, was a more difficult matter, the only possible approach to it being over a plain covered with four hundred yards of abattis, and commanded by ten guns. The British, though they had not a single gun, advanced under a heavy fire, pushed each other through the embrasures, and, in spite of a gallant resistance on the part of the Americans, drove them out of the fort. The American loss was 300 killed, wounded, and prisoners; the British loss, 140 killed and wounded. Having destroyed the American shipping and some other batteries farther up the river, Clinton's little expedition returned to New York. The troop of the Seventeenth formed part of the column that stormed Fort Clinton—a service which, if the original plan of campaign had been

6th Oct.

adhered to, would have been one of the most valuable in the war. 1777.

With this the campaign of 1777 came to an end, decidedly to the disadvantage of the British, who had lost the whole of Burgoyne's division and gained nothing but Philadelphia. The winter of 1777-78 the British army spent in the city of Philadelphia, where it was kept inactive, and allowed to grow slack in discipline and efficiency ; and this although Washington lay for five whole months but 26 miles distant, at Valley Forge—his position weak, his guns frozen into the entrenchments, his army worn to a shadow by sickness and desertion, and absolutely destitute of clothing, stores, and equipment. Howe had 14,000 men, and Washington a bare 4000, yet for the fourth time Howe allowed him to escape ; and this time inaction was fatal, for the new year was to bring with it an event which changed the whole aspect and conduct of operations.

1778. In February 1778 the French Government, still smarting under the loss of Canada, concluded a treaty of defensive alliance with the young American Republic, and despatched a fleet under D'Estaing to operate on the American coast. The British Government no sooner heard the news than it sent instructions for the army to evacuate Philadelphia and retire to New York, from whence half of it was to be forthwith despatched to attack the French possessions in the West Indies. The burden of this duty fell, not upon Howe, to whom it would have been a just retribution, but upon Clinton, who succeeded to the command on Howe's resignation in the spring of 1778.

During the winter the Seventeenth had been moved down from New York to join the main army at Philadelphia, where, in March 1778, we find them reduced to a nominal total of 363 men, of whom no fewer than 67 were in hospital, and 162 horses. Fortunately for its own sake the regiment was busily employed during the spring in the duty of opening communications and bringing in supplies, by which it was prepared for the heavy work that lay before it. On the 3rd of May a strong detachment of the

1778. Seventeenth formed part of a mixed force of 1000 men which was sent out to reduce a hostile post at Crooked Billet, seventeen miles from Philadelphia. The business was neatly managed, for the British, with trifling loss, killed, wounded, or captured 150 of the Americans, and, thanks to the Seventeenth, took the whole of their baggage. Three weeks later the regiment was again employed in a small expedition against 3000 Americans, who had been posted by Washington in an advanced and isolated position at Barren Hill under the command of Marquis Lafayette. This time the affair was sadly bungled, and the Americans, who should have been captured in a body, would have got off scot free but for a dash made on the rear-guard by the Light Dragoons, wherein 40 or 50 American prisoners were taken.

By constant excursions of this kind, on a larger or smaller scale, the regiment was prepared for the very arduous duty that lay before it. On the 18th June, at 3 A.M., the evacuation of Philadelphia was begun, and by 10 A.M. the whole British army had crossed the Delaware at the point of its junction with the Schuylkill. It then advanced up the left bank, on a road running parallel to the river, as far as Cornell's Ferry, where it left the line of the Delaware and turned off on the road to Sandy Hook. Up to the 27th June the British, though constantly watched by small parties of the enemy, were allowed to pursue their march through this difficult country without molestation; but on that day an advanced corps of 5000 Americans appeared close in rear, with the main army of Washington but three miles behind it, while other smaller bodies came up on each flank. On

28th June. the 28th, Clinton, expecting an attack, divided his army into two parts, the first of which he sent off at daybreak in charge of the baggage (which was so abundant that the column was twelve miles long), leading off the second, under his personal command, at 8 A.M. The Seventeenth was attached to the baggage column, and must have marched with it for some eight or nine hours, when it was hurriedly sent for to join the rear-guard under General Clinton. The rear column had just come down from the high ground into a plain about three miles long by one mile wide, when the Americans appeared in force in the rear and on both flanks. Their first attempt was made on the right flank, and was likely to have been serious, had it not been checked, to use Clinton's words, by the resolute bearing and firm front of the Seventeenth. The Americans had not lost their respect for the Light Dragoons. From that point the regiment was swiftly moved to others; and the general impression left on the mind by Clinton's rather confused description is, that the Seventeenth were kept manœuvring round the column, frequently under Clinton's immediate direction, wherever the Americans threatened most danger. The 16th Light Dragoons, more fortunate than the Seventeenth, had a chance of charging the American cavalry, and made admirable use of it; but they lost a great number of horses, which was a serious matter considering the weakness of the British mounted force. Finally Clinton made his dispositions for a pitched battle in the plain; but the Americans knew better than to accept it, and retired to the hills from which they had originally come down. Clinton thereupon attacked them with the infantry and drove them back. They retreated to a second position. Again Clinton attacked, and after hard fighting forced them out. They then fell back on a third position, where, Clinton feeling by this time assured of the safety of his baggage, thought best to leave them. And so ended the very hard day's work which takes its name from the heights of Freehold, at the foot whereof the combat was fought. So terrible was the heat in the confinement of the valley that fifty-nine of the infantry dropped dead while advancing to the attack. The total loss on the English side was 358 men. The Seventeenth had no casualties, though Clinton's testimony shows that they did good work. The Americans lost 361 men, and from that day abandoned the pursuit, having had for the present enough of it. Clinton, therefore, made the rest of his way untroubled to Sandy Hook, and on the 5th July embarked his army for New York. A flying expedition to Rhode Island, which arrived too late to catch

1778.

1778. the French force that had threatened it, and a successful inroad into Georgia in the south, brought the campaign of 1778 to a close.

In November, Clinton, in obedience to his orders, sent away half of his army to England and the West Indies. He was so sensible of the injury inflicted on his forces by the loss of some of his best troops, that he begged to be allowed to resign his command, and required some pressure to induce him to retain it. His difficulties were great enough, for everything was going wrong in New York. In December there was not a fortnight's flour in store, and not a penny in the military chest. The clothing provided for the men proved to be bad, and was condemned right and left by their officers. "The linen is coarse and thin, and unfit for soldiers' shirts, the stockings of so flimsy a texture as to be of little service, and the shoes of the worst kind." One consignment of shoes was found to consist of "thin dancing pumps," and even these too small for the men to wear. Moreover the Government in England, which had always given Howe a free hand, thought it right to tie down Clinton, who was far the better man, with every kind of order. "For God's sake, my Lord," the General wrote at last, "if you wish me to do anything leave me to myself."

Such was the state of things when the Seventeenth went into their winter quarters at Hampstead, Long Island, in 1778. It was now the only British cavalry corps on the American Continent, the 16th having gone home, leaving all its horses and a certain number of men with the sister regiment. Though its numbers were thus raised to 414 men, we shall not again find it in the field entire during the remainder of the war. From this winter onward the scene of the main contest shifts from the north to the south, and we shall find the Seventeenth divided between these two points of the compass.

CHAPTER V

THE AMERICAN WAR—2ND STAGE—THE SOUTHERN CAMPAIGN, 1780-1782

THE alliance of France with the revolted provinces having compelled the British Government to reduce General Clinton's army by one-half, this loss was supplemented by the enlistment of volunteers from the loyal party in America itself, and by the organisation of corps of irregulars. One such corps, consisting partly of cavalry and partly of infantry, was commanded by Captain Lord Cathcart of the Seventeenth, and another, known as the King's American Dragoons, received an Adjutant from the regiment. But the corps with which the name of the Seventeenth was inseparably connected was the so-called "Legion" commanded by Colonel Banastre Tarleton. To this last a small party of the Seventeenth seems to have been permanently attached, probably as a pattern for the guidance of the provincial recruits. But in addition to these a troop of the regiment under its own officers was frequently joined to it, which though in contemporary accounts generally included in the term "Cavalry of the legion," was distinct from it and careful to preserve its individuality.

1780.

With the change in the composition of the army came simultaneously a change in the plan of campaign, by a return to the scheme, already tried once at the outbreak of the war, of an expedition to the Carolinas; where it was hoped that the loyalists were numerous and ready to rally round the army. The plan was to scour the country with flying columns, which would serve at once to hearten good subjects and overawe the

1780. disaffected. For such operations Charleston was required as a base, and it was to preparations for the reduction of Charleston that most of Clinton's energies were devoted in the summer of 1779. An accession of strength was gained by the evacuation of Rhode Island in October, and finally, on the 26th December, Clinton sailed with a portion of his army on this expedition to the South. One troop of the Seventeenth, sixty strong, accompanied him.

Bad luck dogged this enterprise from the first. The transports were overtaken by a storm and dispersed in all directions. All the cavalry horses perished, and one ship containing siege artillery was lost. It was not till the end of January that the ships, many of them badly battered, appeared at the appointed rendezvous, the Island of Tybee, off the coast of Georgia, having spent five weeks over a voyage generally reckoned to last ten days. The troop of the Seventeenth was sent with Tarleton's legion to Port Royal, a little to the north of Savannah, where it was landed and quartered at Beaufort, at the head of the harbour. With great difficulty it procured forty or fifty inferior horses; and after a time was ordered to join some reinforcements that were marching up from Savannah, and advance up country with them to unite with Clinton's army before Charleston. Meanwhile the people of the country, knowing that the British had lost their horses, equipped themselves as cavalry to harass the column on the march. Nothing could have suited Tarleton better. A charge by the troop of the Seventeenth sufficed to disperse these irregular horsemen, and ensure the capture not only of several prisoners, but, better still, of their horses. After twelve days' march through a difficult country broken up by flooded rivers, and in the thick of a hostile population, the legion arrived at its destination on the Ashley with its strength in horses multiplied by four or five, and a good supply of forage to boot.

Meanwhile General Clinton with the rest of the army had sailed to the river Edisto, a little to the south of Charleston, and advanced thence by slow marches upon the town. Charleston lies on a tongue of land which runs, roughly speaking, from north to south, being enclosed between the Cooper River on the east and the Ashley on the west. The British fleet having moved up to blockade it to the south or seaward, Clinton on the 30th March threw his army across the Ashley to the neck of the isthmus on which the town stands, and encamped over against the American entrenchments. As usual these were formidable enough, stretching across the isthmus from the Ashley to the Cooper, and strengthened by a deep canal, two rows of abattis, and other obstacles. Over and above the garrison of 6000 men within the town, the Americans kept a force of militia and three regiments of cavalry, under General Huger, on the upper forks and passes of the Cooper, whereby the communications between the town and the back country were kept open. The dislodgment of this corps of Huger's was therefore indispensable to the complete investment of Charleston ; and the execution of this task was intrusted to a picked force of 1400 men, including Tarleton's legion and the detachment of the Seventeenth.

1780.

On the 12th April, therefore, Tarleton moved off to Goose Creek on his way to Monk's Corner, thirty miles from Charleston, where there lay the American post that held Biggin's Bridge over the Cooper. Knowing that the enemy was superior to him in cavalry, he had determined to make a night attack, and he had the good fortune on the way to pick up a negro who acquainted him with the enemy's dispositions. Learning from this source that the American force was divided, the cavalry being on his own side of the river and the infantry on the other, he pushed on through the night, and at 3 A.M. surprised the main guard of the cavalry. Galloping hard on the backs of the fugitives he dashed straight into the camp, dispersed the far superior force that lay there, and captured 150 prisoners, 400 horses, and 50 ammunition waggons. The bridge being thus uncovered he at once ordered his infantry across it against the American post on the other side ; and this having been captured, detached a force to seize Bowman's Ferry, which commanded another branch of the Cooper. This was promptly done, and by the evening

1780. the American communications on the Cooper were cut through and Charleston completely isolated.

The Americans, however, were not so easily to be baulked. Huger himself and his principal officer, Colonel Washington,[1] had managed to escape by hiding in a swamp, and before the end of April had begun to collect another force of cavalry to the north of the Santee, a river which runs parallel to the Cooper, and at its nearest point is not above twenty miles from Biggin's Bridge. On the 6th of May this force crossed the Santee, snapped up a British foraging party, and prepared to recross the river, a few miles lower down, at Lanew's Ferry. Tarleton, who was patrolling with the detachment of the Seventeenth and some of his own dragoons, 150 men all told, learned what had happened, and pressed on with all haste to catch the Americans before they could repass the Santee. Once again he caught a superior force by surprise. Coming up at 3 P.M. with the American vedettes he at once drove them in upon the picquet, and was on the backs of the main body in an instant. Five officers and 36 men were cut down, 7 officers and 60 men made prisoners, and the rest, including Colonel Washington, driven into the river to escape as best they could by swimming. Tarleton, who had lost but two men and four horses killed, marched back to camp, twenty-six miles, on the same evening, with the result that twenty horses died of fatigue. But Tarleton, as we shall see, never spared men or horses.

On the 12th May Charleston surrendered to General Clinton, who thereupon prepared to return to New York. But first he sent three expeditions up three different rivers to the interior to pursue the advantages gained by the surrender. Of these three, one, under Lord Cornwallis, was ordered to cross the Santee River and pursue a large train of American stores and ammunition which, under the command of Colonel Burford, was retreating in all haste by the north-east bank towards North Carolina. Accordingly, on the 18th May, Cornwallis with a mixed force

[1] This Colonel Washington must not be confounded with his namesake the famous George.

of 2500 men, including Tarleton's legion and the Seventeenth, marched off and crossed the Santee in boats at Lanew's Ferry. The legion and Seventeenth were then at once detached to Georgetown to clear the left flank of Cornwallis's line of march, while the main body pursued its way up the river to Nelson's Ferry. Having rejoined Cornwallis at that point on the 27th, Tarleton was detached once more with 40 men of the Seventeenth, 130 of the legion dragoons, 100 mounted infantry, and a three-pounder field-gun, to follow Burford by forced marches. So intense was the heat that many both of the men and of the horses broke down; but by dint of impressing fresh horses on the road the little column reached Camden (sixty miles distant as the crow flies) on the following day. There Tarleton learned that Burford was still far ahead of him, having left Rugeley's Mills (twenty miles as the crow flies beyond Camden) on the 26th. Moreover, American reinforcements were on the march to join him from North Carolina, and both columns were making all haste to effect a junction. Seeing that such junction must at all hazards be prevented, Tarleton started off again at 2 A.M. on the 29th, reached Rugeley's Mills at daylight, and there obtained information of Burford still in retreat twenty miles ahead of him. In the hope of delaying him Tarleton sent him a message, wherein he exaggerated the strength of his force, to summon him to surrender. But Burford was too cunning either to pause or to surrender; so there was nothing for Tarleton to do but to leave his three-pounder behind and press on with his weary men and horses as best he could. At last at three in the afternoon the British advanced parties came up with Burford's rear-guard, captured five men, and forced Burford to turn and fight. His force was 380 infantry, a detachment of cavalry, and 2 guns. The British had started but 300 strong, had marched a hundred and five miles in fifty-four hours, and had perforce left some men behind them on the way. Tarleton divided his little party into three columns, whereof the men of the Seventeenth, under Captain Talbot, formed the centre, and attacked at once.

1780.

29th May.

1780. The Americans reserved their fire till the cavalry was within ten yards of them, but failed to check the charge of the British, who galloped straight into the middle of them and did fearful execution. Tarleton's horse was killed under him; and the men, thinking that their leader was dead, became mad. The Americans lost 14 officers and 99 men killed; 8 officers and 142 men wounded, 3 officers and 50 men prisoners, also 3 colours, 2 guns, and the whole of their baggage train. The British lost but 2 officers and 3 men killed, 1 officer (Lieutenant Patteshall of the Seventeenth) and 11 men wounded, and 40 horses. After this action, known as the engagement of Waxhaws, the Americans who were advancing from North Carolina at once retired; and Tarleton rejoined Cornwallis at Camden. South Carolina was now virtually cleared of American troops; and Cornwallis having established a few outlying posts to keep order, and left Lord Rawdon in command at Camden, returned to Charleston to take up the business of civil administration.

General Washington now detached 2000 men from the North to North Carolina, which nucleus being reinforced by 4000 more men from Virginia, entered South Carolina once more on the 27th July, and advanced along the line of the Upper Santee upon Camden. To the great disgust and disappointment of the British commander the whole country welcomed the arrival of the Americans with joy, and Cornwallis in great anxiety hastened up to Camden in person. General Gates with 6000 men was advancing in his front, General Sumpter with 1000 men was threatening his communications with Charleston in rear; 800 of the garrison of Camden were in hospital, and a bare 2000 men fit for service. Nevertheless Cornwallis decided rather to advance against Gates than to retreat upon Charleston; and accordingly marched at 10 P.M. on the 15th August, almost exactly at the time when Gates started down the same road to meet him. At

16th Aug. 2 A.M. the advanced parties of the two columns met, fortunately just at a point where Cornwallis had reached a good position, his flanks being secured by swampy ground, and the line of

Gates's advance narrowed by the same cause to a point which prevented deployment of his far superior force. Cornwallis drew up his little army in two lines, holding Tarleton's cavalry in reserve in the rear. Even this small force of mounted men had been weakened by the recall of part of the Seventeenth to New York; but the regiment was nevertheless represented. Cornwallis took the initiative, and after an hour's hard fighting broke up the Americans completely. Then Tarleton was let loose with his men of the Seventeenth and dragoons of the legion, who pursued the defeated army for twenty-two miles, capturing seven guns, the whole of the baggage, and a great number of prisoners. Cornwallis lost 345 men killed and wounded, nearly all of them from the infantry, while the Americans lost in killed, wounded, and prisoners, not far from 2000 men, a number equal to that of the whole British force engaged.

1780.

There still remained General Sumpter, with 1000 men well armed and equipped, on the south side of the Wateree (Upper Santee), who was now preparing to retreat to North Carolina. Tarleton with a mixed force of 350 men was at once sent across the river after him; but by noon on the day after the battle his troops were so exhausted by fatigue and by the heat that he was forced to pick out 100 cavalry and 60 infantry, and proceed with these alone. After marching five miles further his advanced party came upon two American vedettes, who fired and killed one dragoon. But the shots caused no alarm in the American camp, for it was assumed that the American militia-men, according to their usual habit, were merely shooting at cattle. Tarleton's men at once captured the vedettes, and moved on to a neighbouring height, from which on peering over the crest they discovered the Americans comfortably resting, without the least suspicion of danger, during the heat of the day. General Sumpter was not even dressed, so hot was the weather; and altogether Tarleton's task, thanks to his own energy, was once more an easy one. The Americans were promptly attacked and dispersed with the loss of 150 killed and wounded, and 300

17th Aug.

1780. prisoners. Two guns, a great quantity of stores and ammunition, and 250 loyalist prisoners previously captured by Sumpter, also fell into Tarleton's hands.

Emboldened by this success, Lord Cornwallis advanced into North Carolina, but owing to the destruction of one of his detachments was compelled to fall back once more into South Carolina, and thus, notwithstanding his victory at Camden, found himself in as bad a position as ever. In November the indefatigable Sumpter, undismayed by previous defeats, collected another force and again threatened the British communications between Camden and Charleston. Once again Tarleton was ordered to checkmate him; but this time fortune sided with Sumpter. Tarleton on receiving his instructions moved off with his usual swiftness, and interposing between Sumpter's force and the line of retreat into North Carolina, was on the point of cutting him off before Sumpter had received the least warning of an enemy's approach. Unluckily, however, a deserter betrayed Tarleton's movements, and thus enabled Sumpter to get the start of him on his retreat. Tarleton none the less followed hard after him, and having overtaken his rear-guard, and cut it to pieces, hurried forward with a handful of 170 of the Seventeenth and legion cavalry, and 80 mounted infantry, to catch the main body before it could cross a rapid river, the Tyger, that barred its line of march. At 5 P.M. on the 20th November he finally overtook Sumpter at Blackstocks, and with his usual impetuosity attacked him forthwith. The American force was 1000 strong, skilfully posted on difficult ground, and sheltered by log huts. Tarleton's men were beaten back from all points, and being very heavily punished, were forced to retire. But by chance Sumpter himself had been badly wounded; and the Americans, without a leader to hold them together, retreated and dispersed. Tarleton, therefore, although defeated, was successful in gaining his point, and received particular commendation for this action from Lord Cornwallis.

In December reinforcements from New York were sent to

South Carolina, and among them a troop of the Seventeenth, 1780. which was added to Tarleton's command for the forthcoming operations. Cornwallis designed to march once more into North Carolina. The Americans, true to their habitual tactics, resolved to keep him in the South by harassing his outlying posts, and to this end sent 1000 men under General Morgan across the Broad River to attack Lord Rawdon in the district known as "Ninety-six," on the western frontier of South Carolina. Cornwallis replied to this by detaching Tarleton, with a mixed force of about 1000 men, to the north-west to cut off Morgan's retreat. On the night of the 6th January, Tarleton, after a 1781. very fatiguing march, managed to get within six miles of Morgan, who retreated in a hurry, leaving his provisions half-cooked on the ground. At three next morning Tarleton resumed the 7th Jan. pursuit, and at 8 A.M. came up with the American force, disposed for action, at a place called the Cowpens. As usual Tarleton attacked without hesitation, in fact so quickly that he barely allowed time for his troops to take up their allotted positions. The 7th Foot and legion infantry formed his first line, flanked on each side by a troop of cavalry; the 71st Foot and remainder of the cavalry were held in reserve. The Americans were drawn up in two lines, whereof the first was easily broken, but the second stood firm and fought hard. Seeing that his infantry attack was failing, Tarleton ordered the troop of cavalry on the right flank to charge, which it duly did under a very heavy fire, but being unsupported, was driven back by Morgan's cavalry with some loss. Tarleton then ordered up the 71st, which drove back the Americans brilliantly for a time, but being, like the rest of the British force, fatigued by the previous hours of hard marching, could not push the attack home. The Americans rallied and charged in their turn, and the British began to waver. Tarleton ordered his irregular cavalry to charge, but they would not move; and then the American cavalry came down upon the infantry, and all was confusion. "Where is now the boasting Tarleton?" shouted Colonel Washington, as he galloped down on the broken

1781. ranks. But the boasting Tarleton, who had driven Washington once to hide for his life in a swamp, and once to swim for his life across the Santee, was not quite done with yet. Amid all the confusion the troop of the Seventeenth rallied by itself, and with these, a mere 40 men, and 14 mounted officers who had formed on them, Tarleton made a desperate charge against the whole of Washington's cavalry, hurled it back, and pressing on through them, cut to pieces the guard stationed over the captured English baggage. Cornet Patterson of the Seventeenth, maddened by Colonel Washington's taunt, singled him out, and was shot dead by Washington's orderly trumpeter. Lieutenant Nettles of the Seventeenth was wounded, and many troopers of the regiment likewise fell that day. The survivors of that charge were the only men that left the field with Tarleton that evening. The irregular cavalry was collected in the course of the following days; but the infantry men were cut down where they stood. Both the 7th and the 71st had done admirably throughout their previous engagements in the war, and felt that their detachments had not received fair treatment at Cowpens. The 71st, it is on record, never forgave Tarleton to the last.

In spite of his victory Morgan continued his retreat into North Carolina, Lord Cornwallis following hard at his heels, but sadly embarrassed by the loss of his light troops. Having been misled by false reports as to the difficulty of passing the rivers of North Carolina, Cornwallis marched into the extreme back country of the province so as to cross the waters at their head, and on the 1st February fought a brilliant little action to force the passage of the Catawba. At the close of the day Tarleton's cavalry had an opportunity of taking revenge for Cowpens, and this time did not leave the Seventeenth to do all the work alone. From the Catawba Cornwallis pressed the pursuit of Morgan with increased energy, but failed, though only by a hair's breadth, to overtake him. Nevertheless, by the time he had reached Hillsborough, the American troops had fairly evacuated North Carolina; and Cornwallis seized the opportunity to issue a proclamation summoning the loyalists of the province to the royal standard. The Americans replied by sending General Greene with a greatly augmented force back into Carolina. Thereupon the supposed loyalists at once joined Greene, who was thus able to press Cornwallis back to a position on the Deep River. On the 14th March, Cornwallis, always ready with bold measures, marched out with 2000 British to attack Greene with 7000 Americans, met him at a place called Guildford, and defeated him with heavy loss. The cavalry had no chance, though the Seventeenth was present at the action ; but the British infantry was terribly punished : 542 men were killed and wounded in the fight ; and Cornwallis thus weakened was obliged to retire slowly down the river to Wilmington, which he reached on the 7th April.

1781.

The memory of Cornwallis's campaigns in the Carolinas has utterly perished. But although they issued ultimately in failure, they remain among the finest performances of the British rank and file. The march in pursuit of Morgan, which culminated in the action of Guildford and the retreat to Wilmington, alone covered 600 miles over a most difficult country. The men had no tents nor other protection against the climate, and very often no provisions. Day after day they had to ford large rivers and numberless creeks, which (to use Cornwallis's own words), in any other country in the world would be reckoned large rivers. When, for instance, the Guards forced the passage of the Catawba, they had to ford a rapid stream waist-deep for five hundred yards under a heavy fire to which they were unable to reply. The cavalry on their part came in for some of the hardest of the work, being continually urged on and on to the front in pursuit of an enemy which they could sometimes overtake, but never force to fight ; constantly engaged in petty skirmishes, losing a man here and a man there, but gaining little for their pains, and at each day's close driven to their wits' end to procure food for themselves and forage for their horses. By the time Cornwallis reached Wilmington the cavalry were about

1782. worn out with their work on the rear-guard, and, in Cornwallis's words, were in want of everything. But not a man of the army complained, and all, by Cornwallis's own testimony, showed exemplary patience and spirit. Meanwhile the Americans gave him no rest. No sooner was his back turned on South Carolina than they attacked his posts right and left, making particular efforts against Lord Rawdon at Camden. In fact, in spite of all the hard work done and the hardships endured with invincible patience by the British troops, the state of the country was worse than ever—armed parties of Americans everywhere and all communications cut. Cornwallis was painfully embarrassed by his situation. To re-enter South Carolina would be to admit that the operations of the past eighteen months had been fruitless. He decided that the best course for him was to continue his advance into Virginia, at the same time despatching messengers to warn Lord Rawdon that he must prepare to be hard beset.

Not one of these messengers ever reached Lord Rawdon. The perils of bearers of despatches at this time were such that they could only be conquered by more than ordinary devotion to duty. Fortunately an instance of such devotion has been preserved for us from the ranks of the Seventeenth. The case is that of a corporal, O'Lavery by name, who was especially selected to accompany a bearer of despatches on a dangerous and important mission. The two had not gone far before they were attacked, and both of them severely wounded. The man in charge of the despatch died on the road; the corporal took the packet from the dead man's hand and rode on. Then he too dropped on the road from loss of blood, but sooner than suffer the papers to fall into the hands of the enemy, he concealed it by thrusting it into his wound. All night he lay where he fell, and on the following morning was found alive, but unable to do more than point to the ghastly hiding-place of the despatch. The wound thus maltreated proved to be mortal, and Corporal O'Lavery was soon past all human reward. But Lord Rawdon, unwilling that such gallant service should be forgotten, erected a monument to O'Lavery's memory in his native County Down. 1782.

On the 25th of April Cornwallis, having refreshed his army, quitted Wilmington and marched northward to Petersburg, where he effected a junction with two bodies, amounting together to 3600 men, which had been despatched to reinforce him from England and New York. With these he crossed the Appomattox in search of Lafayette, and pursued him for some way north, destroying all the enemy's stores as he went. The Americans were now, in spite of their continued resistance in South Carolina, in a distressed and desponding position; but just at this critical moment their hopes were revived by intelligence of coming aid from France. Clinton having discovered this by interception of despatches, and learned further that an attack on New York was intended, recalled half of Cornwallis's troops to his own command, and thus put an end to further operations in the South. It is significant that Clinton begs in particular for the return of the detachment of the Seventeenth; evidently he counted upon this regiment above others in critical times. Thus for the moment operations in the South came to a standstill, and Cornwallis retired to Yorktown. 20th May.

Meanwhile Washington had raised an army in Connecticut and marched down with it to his old position at Whiteplains, where he was joined by a French force of 6000 men which had occupied Rhode Island since June of the previous year. For more than a month Washington kept Clinton in perpetual fear of an attack, until at last he received intelligence that the expected French fleet under the Comte de Grasse was on its way to the Chesapeake. Then he suddenly marched with the whole army, French and American, to Philadelphia, and thence down the Elk River to the Chesapeake. De Grasse had been there with 24 ships and 3500 troops since the 30th, and had managed to keep his position against the British fleet of 19 ships under Admiral Graves. This brief command of the sea by the French virtually decided the war. Yorktown was invested on the 28th September,

1782. and on the 19th October Cornwallis was compelled to surrender. From that moment the war was practically over, though it was not until the 16th April 1783 that Washington received, from the hand of Captain Stapleton of the Seventeenth, the despatch that announced to him the final cessation of hostilities.

So ended the first war service of the 17th Light Dragoons. It will have been remarked that since 1779 little has been said of the headquarters of the regiment stationed at New York. The answer is that there is little or nothing to say, no operations of any importance having been undertaken in the North after the capture of Charleston. Yet it is certain that the duties of foraging, patrolling, and reconnaissance must have kept the men in New York perpetually engaged in trifling skirmishes and petty actions, whereof all record has naturally perished. A single anecdote of one such little affair has survived, and is worth insertion, as exemplifying from early days a distinctive trait of the regiment, viz. the decided ability of its non-commissioned officers when left in independent command. We shall find instances thereof all through the regiment's history. Our present business is with Sergeant Thomas Tucker, who, when out patrolling one day with twelve men, came upon a small American post, promptly attacked it, and made the garrison, which, though not large, was larger than his own party, his prisoners. Tucker had accompanied the regiment from England as a volunteer; he went back with it to England as a cornet. Incidents of this kind must have been frequent round New York; and as seventeen men of the Seventeenth, exclusive of those taken at Yorktown, were prisoners in the hands of the Americans at the close of the war, there can be no doubt that the garrison duty in that city was not mere ordinary routine.

A few odd facts remain to be noted respecting the officers. The first of these, gleaned from General Clinton's letter-book of 1780, is rather pathetic. It consists of a memorial to the King from the 17th Light Dragoons, setting forth "that they look upon themselves as particularly distinguished, by having been employed in the actual service of their country ever since the rebellion began in America. But its being the only regiment of Dragoons in this service, and their promotion being entirely confined to that line, they cannot but feel sensibly when they see every day promotion made over them of officers of inferior rank." I cannot discover that the least notice was taken of this petition, hard though the case undoubtedly was; for many of these officers held high staff appointments in New York. Lieutenant-Colonel Birch was a local Brigadier-General, and towards the end of the war was actually in command at New York; but he seems to have gained little by it. On the other hand Captain Oliver Delancey made his fortune, professionally speaking, through his success as Clinton's Adjutant-General from August 1781.

1782.

As to the detachments employed in the South enough has already been said. But it is worth while to correct the error into which other writers have fallen, that the men of the Seventeenth were not with Cornwallis in the campaign of North Carolina. The fact is rendered certain by the mention of twenty-five men in the melancholy roll of the capitulation of Yorktown, which twenty-five I take to be the remnant of the small body that was permanently attached to Tarleton's legion. Moreover, it was not likely that Cornwallis, who was badly in want of light troops, would have left them to do garrison work with Rawdon. The loose expression "legion-cavalry" is so often used to cover the whole of the mounted force under Tarleton's command, that it is frequently difficult to distinguish the detachment of the Seventeenth from the irregulars. But the men of that detachment were not willing to sink their individuality in the general body of legion dragoons. When their old regimental uniform was worn out they were offered the green uniform of the legion, but they would have none of it. They preferred to patch their own ragged and faded scarlet, and be men of the Seventeenth. Nor can we be surprised at it when we remember how the legion retired and left a handful of the Seventeenth to face the victorious Americans alone at Cowpens. This action gives a fair clue to the real seat of strength in Tarleton's cavalry.

1782. Lastly, it must be noted that, although the history of the American War is usually slurred over in consequence of its disastrous conclusion, yet to the rank and file of the British army there is far more ground therein for pride than for shame. British troops have never known harder times, harder work, nor harder fighting, than in the fifteen hundred miles of the march through the Carolinas. They were continually matched against heavy odds under disadvantageous conditions, yet they were almost uniformly victorious. The Americans fought and kept on fighting with indomitable courage and determination, but it was not the Americans but the French, and not so much the French army as the French fleet, that caused Cornwallis to capitulate at Yorktown.

CHAPTER VI

RETURN OF THE 17TH FROM AMERICA, 1783—IRELAND, 1793—EMBARKATION FOR THE WEST INDIES, 1795

IN 1783 the Seventeenth embarked from New York and returned to Ireland, after an absence of eight years. I have failed to discover the exact date. The last muster in America is dated New York, 29th June 1783; the first in Ireland, Cork, 14th January 1784, which latter date must be approximately that of their arrival. This muster-roll at Cork is somewhat of a curiosity. Firstly, it is written on printed forms, the earliest instance thereof in the history of the Seventeenth; in the second place, it shows the regiment to be 327 men short of its proper strength, which is, to say the least of it, singular; and, lastly, it shows that every troop had lost exactly forty horses, no more and no less, cast and dead in America,—a coincidence which sets one wondering who may have been the person or persons that made money out of it. The regiment was now reduced to a peace establishment of 204 non-commissioned officers and men, and stationed at Mount Mellick, Maryborough, and other quarters in King's and Queen's Counties. It also received new clothing, and for the first time discarded the scarlet, which it had hitherto worn, for blue.

1783.

1784.

The new kit, which, saving regimental distinctions, was issued to the whole of the Light Dragoons, consisted of a blue jacket, with white collar and cuffs and the whole front laced with white cord, similar to the jackets now worn by the Horse Artillery. The shade of blue was dark for regiments serving at

April.

F

1784. home, and French gray for regiments serving in India. The helmet also was altered to the new and seemingly very becoming pattern which is to be seen in so many old prints. The leather breeches remained the same, but the boots, for officers at any rate, were more in the Hessian style. A coloured picture published at the beginning of the century makes the new dress appear a very handsome one, in the case of the Seventeenth Light Dragoons—the combination of light blue, silver lace, and crimson sash, relieved by the black fur on the cap, being decidedly pleasing. Let us note that the Seventeenth still retained their mourning lace round the helmet, and the plume of scarlet and white. The badge, of course, appears both on helmet and sabre-tasche, though, if so small a point be worth notice, the skull is below and not above the cross-bones. Shoulder-belts continued to be of buff leather, but the sword-belt of 1784, henceforward worn round the waist, was black. It is painful to have to add that in this year, when the Light Dragoons were on the whole more becomingly and sensibly dressed than at any other period of their existence, the abomination known as the shako made its first appearance in the cavalry, being in fact the head-dress for field-day order. Though not yet quite so extravagantly hideous as it became under King George IV. it was sufficiently ugly—felt in material and black in colour, with white lace curling spirally around it, and a short red and white plume.

Of the life of the regiment during the nine ensuing years there is neither material nor, I think, occasion for an annual chronicle. Lieutenant-Colonel Samuel Birch still retained the command, and held it until 1794. The only one of the original officers that remained, Captain Robert Archdale, disappears from the regimental list after 1794, so that for two whole years Birch was the sole survivor.

Meanwhile these were troublous days for Ireland. In the course of the American War the country had been so far stripped of troops that, in the alarm of French invasion in

1779, corps of volunteers, to the nominal strength of 50,000 men, had been raised for purposes of defence. Unfortunately, however, these volunteers did not confine themselves to military matters. They were, in Mr. Froude's words, armed politicians not under military law. As such they twice received the thanks of the Irish House of Commons for political services, and finally extorted the independence of the Irish Parliament in 1782. They then attempted to establish a Legislative Assembly side by side with the House of Commons, and virtually to dictate to it the government of the country, and this although the peace of 1783 had rendered their existence as a defending force wholly unnecessary. They were suppressed by a little firmness, and therewith their character changed. Hitherto, though supported in part by Catholic subscriptions, the volunteers had consisted of Protestants only—men of position and good character. These men now retired, and their arms fell into the hands of ruffians and bad characters of every description. At last in 1787 these volunteers, once the idol of Ireland, appeared to have ceased their existence, but it was only for a time.

1784.

The outbreak of the French Revolution in 1789, with its cant words of liberty, equality, and fraternity, turned many heads all the world over, and nowhere more than in Ireland. The most significant symptom thereof was the foundation of the Society of United Irishmen by the rebel Wolfe Tone; whereof the main object was the propagation and adoption of revolutionary principles, and ultimately rebellion. In 1792 some of Tone's associates formed two battalions of "National Guards," which were to hold a great review on the 9th December, but having been informed that they would muster at their peril, very sensibly took care, after all, not to put in an appearance. This happened in Dublin. But at Belfast and in the North there was not less sympathy with the Jacobins and the extreme revolutionists of France, and in Ulster too there were "National Guards" of the same stamp.

1792.

The services of a regiment in aid of the civil power are

1793. so ungrateful that they are better left unrecorded, nor would allusion here be made to those of the Seventeenth but for the coincidence that they have found a place in history. For in the year 1786 began one of those periodic outbreaks of agrarian crime which have so often troubled Ireland, the perpetrators being what are now called moonlighters but were then known as whiteboys or defenders. Of the share taken by the Seventeenth in the suppression of these defenders it is best to say nothing, arduous though the work undoubtedly was. But it was a far more serious matter when, early in April 1793, the "National Guard" of Northern Republicans paraded in their green uniforms at Belfast, undeterred by the suppression of their brethren in Dublin. In March, General Whyte was sent down to compel their submission, the Seventeenth forming part of his force. He thereupon sent four troops of the regiment to disarm the "Guard" of these Republican volunteers. The rest of the story is best told in Mr. Froude's own words :—

1793. On the evening of the 9th March, a corporal and a private of the 17th, off duty, strolled out of the barracks into the city where they met a crowd of people round a fiddler who was playing *Ça ira*. They told the fiddler to play God save the King. The mob damned the King with all his dirty slaves, and threw a shower of stones at them. The two dragoons, joined by a dozen of their comrades, drew their sabres and "drove the town before them." Patriot Belfast had decorated its shops with sign-boards representing Republican notables. The soldiers demolished Dumouriez, demolished Mirabeau, demolished the venerable Franklin. The patriots so brave in debate, so eloquent in banquet, ran before half a dozen Englishmen. A hundred and fifty volunteers came out, but retreated into the Exchange and barricaded themselves. The officers of the 17th came up before any one had been seriously hurt, and recalled the men to their quarters. In the morning General Whyte came in from Carrickfergus, went to the volunteer committee room, and said that unless the gentlemen in the Exchange came out and instantly dispersed, he would order the regiment under arms. They obeyed without a word. The dragoons received a reprimand, but not too severe, as the General felt that they had done more good than harm.[1]

[1] Froude, *English in Ireland*, iii. 105, 106.

Thus through two men of the Seventeenth the Irish volunteers were finally brought to an end. It must be remembered in defence of these two dragoons that their regiment had fought through the whole of the American War, which had failed mainly through the Alliance of the French with the Americans; and that it was a little hard on them, when at home, to hear abuse of the King whom they served, and witness the exaltation of French and American heroes. Moreover, in those days the Irish had injured so many soldiers by hamstringing them when peaceably walking in the streets that there was a deal of bad blood between the Irish and the Army.

1793.

In that same year began the great war with France which was destined to last, with only a few months intermission, for the next twenty years, and to be finally closed by the victory of Waterloo. The efforts of Mr. Pitt were early directed against the French possessions in the West Indies—a policy which, after having been for many years condemned, in deference to the verdict of Lord Macaulay, has lately been vindicated by a more competent and impartial authority, Captain Mahan of the United States Navy. The richest of the French West Indies was the Island of St. Domingo, which accordingly became one of Pitt's first objects. Ever since 1790, when the revolutionary principles of Paris had first found their way thither, the island had been in a state of disturbance, which had culminated, partly through mismanagement and partly through wilful mischief, in a general rising of the negroes against the whites, accompanied by all the atrocities that inevitably attend a servile war and a war of colour. Of the white planters many took refuge in Jamaica, whence they pressed the British Government to take possession of St. Domingo, averring that all classes of the population would welcome British dominion, and that on the first appearance of a British force the Colony would surrender without a struggle. It was the story of the Carolinas repeated, and we shall see that the story had the same end.

St. Domingo, an island almost as large as Great Britain,

1793. in shape greatly resembles a human right hand cut off at the wrist, and with the thumb, second and third fingers doubled inwards; the wrist forming the eastern end, and two long promontories, represented by the little and first fingers, the western extremities. The French garrison in the island consisted of 6000 regular troops, 14,000 white militia, and 25,000 negroes. The British force first directed against it consisted of 870 rank and file, which with the help of a small squadron captured and garrisoned the ports of Jeremie and Mole St. Nicholas, situated near the extremities of the south and north promontories respectively. These posts, as commanding the windward passage between St. Domingo and Cuba, were of considerable strategic importance to the Navy. From Jeremie an expedition was undertaken against Cape Tiburon, in reliance on the help of 500 friendly Frenchmen, whom a French planter undertook to raise for the purpose. Not 50 Frenchmen appeared, and the attack was a total failure. Then came the rainy season, and with it the yellow fever, which played havoc among the troops. Reinforcements being imperatively needed, more men were withdrawn from Jamaica to St. Domingo, whereby, as will presently appear, the safety of Jamaica was seriously compromised.

19th Sept.
22nd Sept.

1794. In the spring of 1794 the British succeeded in taking Tiburon and one or two more ports, and finally in June they effected the capture of Port au Prince. But the revolted negroes, under the command of a man of colour, Andrew Rigaud, showed plainly by an attack on the British post at Tiburon that they at any rate did not mean to accept British rule. And now yellow fever set in again with frightful severity. A small British reinforcement of 300 men lost 100 in the short passage between Guadeloupe and Jamaica, left 150 more dying at Jamaica, and arrived at Port au Prince with a bare 50 fit for duty. Then Rigaud again became active, and on 28th December succeeded in recapturing Tiburon, after the British had lost 300 men out of 480.

1795.

When the news of all these calamities arrived in England, it 1795. was resolved that four regiments of Light Cavalry should be sent dismounted to St. Domingo in August, and that meanwhile detachments amounting to eight troops of the 13th, 17th and 18th Light Dragoons should be despatched to Jamaica forthwith. These last were, if required by the General, to be sent on to St. Domingo; and as the General required them very badly, being able to raise only 500 men fit for duty out of seven regiments, he lost no time in asking for them.

The detachments, including that from the Seventeenth, were accordingly shipped off, when or from whence I have been unable to discover. As little is known of the life on a transport in those days, it may be worth while to put down here such few details as I have succeeded in collecting. In the first place, then, hired transports seem generally to have been thoroughly bad ships. That they should have been small was unavoidable; but they seem as a rule to have been in every respect bad, and by no means invariably seaworthy. Those who have seen in the naval despatches of those days the extraordinary difficulty that was found in keeping even men-of-war clean, and the foul diseases that were rampant in the fleet through the jobbery and mismanagement of the Admiralty, will not be inclined to expect much of the hired transports. Let us then imagine the men brought on board a ship full of foul smells from bad stores and bilge-water, and then proceed to a brief sketch of the regulations.

The first regulation is that the ship is to be frequently fumigated with brimstone, sawdust, or wet gunpowder—no doubt to overcome the pervading stench. Such fumigation was to begin at 7 A.M., when the berths were brought up and aired, and be repeated if possible after each meal. Moreover, lest the free circulation of air should be impeded unnecessarily, it was ordained that married couples should not be allowed to hang up blankets, to make them separate berths, *all over the ship*, but in certain places only. The men were to be divided into

1795. three watches, one of which was always to be on deck; and in fine weather every man was to be on deck all day, and kept in health and strength by shot drill. For the rest the men were required to wash their feet every morning in two tubs of salt water placed in the forecastle for the purpose, to comb their heads every morning with a small tooth comb, to shave, to wash all over, and to put on a clean shirt at least twice a week.

At the very best the prospects of a voyage to the West Indies a century ago could not have been pleasant; but the experience of these unfortunate detachments of dragoons seems to have been appalling. After a terrible passage, in which some ships were cast away, and all were seriously battered, a certain number of transports arrived in July at Jamaica, and among them those containing two troops of the Seventeenth. Jamaica not being their destination, they were told that their arrival was an unfortunate blunder, and packed off again to St. Domingo. Think of the feelings of those unhappy men at being bandied about in such a fashion. They had not sailed clear of the Jamaican coast, however, when they were hastily recalled. The Maroons had broken out into rebellion; and the "unfortunate blunder" which brought the Seventeenth to Jamaica was fated to prove a piece of great good luck to the island and a cause of distinction to the regiment. But something must first be said of the story of the Maroons themselves.

CHAPTER VII

THE MAROON WAR IN JAMAICA, 1795

THE year 1795, as will presently be told when we speak of the services of the Seventeenth in Grenada, was marked by a simultaneous revolt of almost all the possessions of the British in the West Indies. Amid all this trouble the large and important island of Jamaica remained untouched. This was remarkable, for from its wealth it offered a tempting prey to the French, and, from its proximity to St. Domingo, it was easy of access to French agents of sedition and revolt, who could pass into it without suspicion among the hundreds of refugees that had fled from that unhappy island. Moreover, the garrison had been reduced to great weakness by the constant drain of reinforcements for St. Domingo. Still, in spite of some awkward symptoms, the Jamaica planters remained careless and supine; and no one but the governor, Lord Balcarres, a veteran of the American War, felt the slightest anxiety. Such was the state of affairs when the squadron of the Seventeenth arrived at Port Royal in July, and was sent on board ship again. Three days later the Maroons were up in rebellion.

1795.

The history of these Maroons is curious, and must be told at some length if the relation of the war is to be rightly understood. Jamaica was originally gained for the English by an expedition despatched by Cromwell in 1655; but it was not until 1658 that the Spaniards, after a last vain struggle to expel the British garrison, were finally driven from the island. On their departure their slaves fled to the mountains, and there for some

1795. years they lived by the massacre and plunder of British settlers. They seem to have scattered themselves over a large extent of country, and to have kept themselves in at least two distinct bodies, those in the north holding no communication with those in the south. These latter, in their district of Clarendon, being disagreeably near the seat of Government, the British authorities contrived to conciliate and disperse; but their fastnesses had not long been deserted by the Maroons when they were occupied (1690) by a band of revolted slaves. These last soon became extremely formidable and troublesome, their ravages compelling the planters to convert every estate-building into a fortress; and at last the burden of this brigandage became so insupportable that the Government determined to put it down with a strong hand.

At the outset the attacks of the whites on these marauding gangs met with some success; but soon came a new departure. A man of genius arose from among these revolted slaves, one Cudjoe by name, by whose efforts the various wandering bands were welded into a single body, organised on a quasi-military footing, and made twice as formidable as before. Nor was this all. The Maroons of the north, who from the beginning had never left their strongholds nor ceased their depredations, heard the fame of Cudjoe, joined him in large numbers, and enlisted under his banner. Yet another tribe of negroes, distinct in race from both the others, likewise flocked to him; and the whole mass thus united by his genius grew, about the year 1730, to be comprehended, though inaccurately, by the whites under the name of Maroons (hog-hunters). Cudjoe now introduced a very skilful and successful system of warfare, which became traditional among all Maroon chiefs. The grand object was to take up a central position in a "cockpit," *i.e.* a glen enclosed by perpendicular rocks, and accessible only through a narrow defile. A chain of such cockpits runs through the mountains from east to west, communicating by more or less practicable passes one with another. These glens run also in parallel lines from north to south, but the sides are so steep as to be impassable to any but a Maroon. Such were the natural fortresses of these black mountaineers, in a country known to none but themselves. To preserve communication among themselves they had contrived a system of horn-signals so perfect that there was a distinct call by which every individual man could be hailed and summoned. The outlets from these cockpits were so few that the white men could always find a well-beaten track which led them to the mouth of a defile; but beyond the mouth they could not go. A deep fissure, from two hundred to eight hundred yards long, and impassable except in single file, was easily guarded. Warned by the horns of the scouts that an enemy was approaching, the Maroons hid themselves in ambush behind rocks and trees, selected each his man, shot him down, and then vanished to some fresh position. Turn whither he might, the unlucky pursuer was met always by a fresh volley from an invisible foe, who never fired in vain.

Nevertheless the white men were sufficiently persistent in their pursuit of Cudjoe to force him to abandon the Clarendon district; but this only made matters worse, inasmuch as it drove him to an impregnable fastness, whence there was no hope of dislodging him, in the Trelawney district farther to the north-west. This cockpit contained seven acres of fertile land and a spring of water. Its entrance was a defile half a mile long; its rear was barred by a succession of other cockpits, its flanks protected by lofty precipices. Here Cudjoe made his headquarters and laughed at the white men. The Maroons lived in indolent savagery while their provisions lasted, and in active brigandage when their wants forced them to go and plunder. They were fond of blood and barbarity, as is the nature of savages, and never spared a prisoner, black or white. After nine or ten years of successful warfare Cudjoe fairly compelled the whites to make terms with him; and accordingly, in the year 1738, a solemn treaty was concluded between Captains Cudjoe, Johnny, Accompong, Cuffee, Quaco, and the Maroons of Trelawney town on the one part, and George the Second, by the Grace of God King of Great Britain,

1795. France, and Ireland, and of Jamaica Lord, on the other. The terms of the treaty granted the Maroons amnesty, fifteen hundred acres of land, and certain hunting rights; also absolute freedom, independence, and self-government among themselves—the jurisdiction of the chiefs being limited only in respect of the penalty of death, and of disputes in which a white man was concerned. On their part the Maroons undertook to give up runaway slaves, to aid the king against all enemies, domestic and foreign, and to admit two white residents to live with them perpetually. A similar treaty was concluded with another body of Maroons that had not followed Cudjoe to Trelawney from the windward end of the island; and thus the Maroon question for the present was settled.

From 1738 till 1795 the Maroons gave little or no trouble. They remained dispersed in five settlements, three of them to windward, but the two of most importance to leeward, in Trelawney district. They lived in a state midway between civilisation and barbarism, retaining the religion—a religion without worship or ceremony—which their fathers had brought from Africa, cultivating their provision grounds regularly, if in rather a primitive fashion, breeding horses, cattle, and fowls, hunting wild swine and fugitive slaves, and conducting themselves generally in a harmless and not unprofitable manner. Their vices were those of the white man, drinking and gambling, which of course gave rise to quarrels; but they were ruled with a strong hand by their chiefs, and kept well within bounds. Owing to the climate in which they lived, some thousands of feet above the sea, and the free, active life which they led, they were physically a splendid race—tall and muscular, and far superior to the negro slaves whom, from this cause as well as in virtue of their own freedom, they held in great contempt. Moreover, the fact that they were employed to hunt down runaway slaves helped greatly to make them friendly to the whites and hostile to the blacks. In fact they held an untenable position, being bound to the whites by treaty, and fighting in alliance with them both against insurgent negroes, as in 1760, and white invaders, as in 1779-80, and yet bound by affinity of race and colour to the very negroes that they helped to keep in servitude. Meanwhile they grew rapidly in numbers and consideration. Certain restrictions to which they had been subjected by Acts of the Jamaica Assembly at the time of the treaty fell into disuse, and became a dead letter. They began to leave their own district and wander at large about the plantations, making love to the female slaves, becoming fathers of many children by them, and thus gradually breaking down the barrier between themselves and their fellow-blacks. Simultaneously the internal discipline of the Maroons became seriously relaxed. Cudjoe and his immediate successors had ruled them with a rod of iron; but at a distance of two generations the authority of the chiefs, though they still bore the titles of Colonel and Captain, had sunk to a mere name. For a time the Colonel's power in Trelawney was transferred to one of the white residents, a Major James, who had been brought up among the Maroons, could beat the best of them at their feats of activity and skill, and was considered to be almost one of themselves. Of great physical strength and utterly fearless, he would interpose in the thick of a Maroon quarrel, heedless of the whirling cutlasses, knock down those that withstood him, and clap the rebellious in irons without a moment's hesitation. Naturally so strong a man was a great favourite with the Maroons, who, while he remained among them, were kept well in hand. But it so happened that James succeeded to the possession of an estate which obliged him to spend most of his time away from the Maroon town; and as a resident who does not reside could be satisfactory neither to his subjects at Trelawney nor his masters at Kingston, he was deprived of his post. He, rather unreasonably, felt himself much aggrieved by the Government in consequence; and the Maroons, who had been annoyed at his former neglect, became positively angry at his involuntary removal. In plain truth, the Maroons through indiscipline had got what is called "above themselves," and were ripe for any mischief.

1795.

1795. It was not long before matters came to a crisis. The new resident appointed in place of James, though in character irreproachable, was not a man to dominate the Maroons by personal ascendency and courage. A trifling dispute sprang up in the middle of July; the Trelawney Maroons drove him from the town, and on the 18th sent a message to the magistrates to say that they desired nothing but battle, and that if the white men would not come to them and make terms, then they would come down to the white men. With that they called in all their people, and sent the women into the bush—nay, report said that they proposed to kill their cattle and also such of their children as were likely to prove an encumbrance to them.

Lord Balcarres, when the news reached him, was not a little troubled. At ordinary times it might have been politic to temporise and conciliate, but now that the greater number of the islands were aflame such policy seemed impossible. Here was a race of black men in insurrection, who had successfully resisted the whites two generations before, and now held an independent position in virtue of a solemn treaty. The bare existence of such a community was a standing menace at such a time. There was evidence that French agents were at work in Jamaica; and it was remarkable that just at this time the negroes on nine plantations, where the managers were known to be men of unusual clemency, showed symptoms of unrest and discontent. It is evident from Balcarres's despatches that he had negro insurrection, so to speak, on the brain, and it is certain that he was ambitious of military glory; but he cannot be blamed at such a time for acting forcibly and swiftly. For a fortnight endeavours were made to smoothe matters over, and with some slight success, for six of the chiefs surrendered. But the main body still held aloof; and Balcarres without further ado proclaimed martial law. He took pains to obtain information as to every path and track that led into the Maroon district, his plan being to seize these and thus blockade the whole of it, though he admits that it would be a difficult manœuvre to do so effectually "on a circle of forty square miles of the most difficult and mountainous country in the universe." On the 9th August the preparations were complete, and the passes were seized; whereupon thirty-eight of the older and less warlike Maroons surrendered, and were carried away under a guard and kept in strict confinement. Seeing this the remainder at once set fire to their towns (the old and the new town, as the two groups of shanties half a mile apart were named), an action which was not misinterpreted as "a signal of inveterate violence and hostility." It was now clear that the matter would have to be fought out.

1795.

The force at Balcarres's disposal was not great. The garrison consisted of the 16th and 62nd Foot, both so weak as to number but 150 men apiece fit for duty, and the 20th or Jamaica Light Dragoons. Besides these there were the stray detachments of the 13th, 14th, 17th, and 18th Light Dragoons, and of the 83rd Foot, some of them very weak, and probably amounting in all to little more than 400 men. Also there was a fair force of local militia, with several local Major-Generals. The Maroons of Trelawney numbered 660 men, women, and children; and there were at least as many more in the other Maroon settlements, which latter, though they never rose, were greatly distrusted by the Governor. Balcarres resolved to surround the whole of the Trelawney Maroon district, and made his dispositions thus:—Colonel Sandford, with the 16th Foot and 20th Dragoons, covered one outlet to the north; Colonel Hull, with 170 men of the 62nd Foot and of the Seventeenth, another; Colonel Walpole, with 150 of the 13th and 14th Dragoons, barred one approach from the south; and Balcarres himself, with the 83rd, took post to the south-west. The Seventeenth was represented by one troop only, the other being on board ship on its way to St. Domingo.

12th Aug.

On the 12th August the Maroons opened the war by attacking a militia post, and killing and wounding a few men. On the same day Lord Balcarres ordered Colonel Sandford to attack and carry the new town from his side, and having done

1795. so, to halt and cut off the retreat of the Maroons, while he himself attacked the old town from his own side. Off started Colonel Sandford, accordingly, with forty-five of the 18th Dragoons, mounted, a body of militia infantry, and a number of volunteers—the latter men of property in the country, and "all generals," as Balcarres sarcastically remarked. In spite of the steepness and difficulty of the ground the little column advanced rapidly with great keenness. The Maroons on their approach quietly evacuated the site of the new town, and

12th Aug. withdrew into a deep defile, three-quarters of a mile long, which formed their communication with the old town. Presently up came Sandford, and to his great joy carried the new town without opposition. Flushed with success he started off, in disobedience to orders, to take the old town, pressing on with his mounted men, dragoons, and volunteers, at such a pace that the militia could not keep up with him. Thus hurrying into the trap laid for him, he plunged into the defile. The column, which was half as long as the defile, had passed two-thirds of the way through it, when a tremendous volley was poured into its whole length. Not a Maroon was to be seen, and the column continued its advance. A second volley followed: Colonel Sandford fell dead; and then the column began to run. The officer of the 18th, seeing that retreat through the defile would be fatal, dashed straight forward at a small party of Maroons which he saw ahead, broke through them, and galloping headlong through a breakneck country, brought the remains of his detachment safely to Lord Balcarres's camp. Two officers and thirty-five men were killed, and many more wounded in this little affair; and the militia (who had not been under fire) were so far demoralised that they evacuated the new town and retired. That night (though Balcarres knew it not) every Maroon warrior got blind drunk. Sixty of them were so helpless even on the following afternoon that they had to be carried into the cockpit by the women.

Though the Seventeenth was not engaged in this affair, it

has been necessary to describe it at length in order to show how formidable an enemy these Maroons were. Two days after the engagement the second troop of the regiment was disembarked from the transport in Montego Bay, and moved up to the front. British dragoons have rarely been better mounted than these detachments in Jamaica. The island is famous for its horses; and every trooper rode a thoroughbred.

1795.

14th Aug.

Mortified by his failure, Balcarres hurried up reinforcements of militia and stores, the conveyance of the latter proving, from the difficulty of the country, to be a frightful task. On the 18th August he reoccupied the new town, unopposed, and on the 23rd moved with three columns under Colonels Fitch, Incledon, and Hull, against the old town. The march was made at daybreak and in profound silence; and the old town was duly captured, as Balcarres fondly imagined, by surprise. The real fact was that the Maroons, disliking the insecurity of the towns, had evacuated them a week before and withdrawn into the cockpits, leaving only a small alarm-post outside. These Maroon sentries fired a few shots and wounded three men, two of them troopers of the Seventeenth, and quietly retired upon their main body. Balcarres then established a post and a block-house on the site of the new town, occupied every approach, and set himself to destroy all the Maroon provision grounds, with the idea of cooping them up and starving them out. He might as well have tried to pen a swarm of mosquitoes in a lion's cage. The Maroons quietly passed out and burnt and plundered an estate house six miles in rear of Balcarres's headquarters.

At the end of August the rainy season set in, and transport became a matter of extreme difficulty. Balcarres himself returned to Port Royal, and left to Colonel Fitch the duty of completing the cordon round the Maroon district. Fresh obstacles cropped up at every moment. The principal planters to the south-west of the Maroon district, by which side access to it was easiest, were relations of Major James, who took up his

1795. grievances warmly and laid themselves out to thwart the Governor. One of these, a local Major-General, eighty years of age, and recently married to a wife of twenty, took offence because Balcarres appointed a regular Major-General to command the field force over his head. Another local Major-General suddenly abandoned operations with his militia in the middle of a concerted movement, on the remarkable ground that he had promised his wife to return to her in a week, and had already been absent ten days. It was only with the greatest difficulty that the troops, exposed to most arduous service and every possible hardship, could be kept supplied with food. Frequently they passed the whole day without a morsel to eat. To discourage them still further, the militia went home and left the regulars to do all the work; and, finally, the climax came when the commanding officer, Colonel Fitch himself, was caught
12th Sept. in an ambuscade, and with two other officers shot dead.

The control of the operations was now entrusted to Colonel Walpole, who at once hastened to Trelawney with all speed. He found the troops sickly and dispirited, and worn out with incessant duty. It was pretty clear that the idea of confining the Maroons by a cordon was an absurdity, and that the destruction of their provision ground only drove them oftener afield to massacre, plunder, and destroy. After weeks of hard work the small British force had lost two field officers and seventy men killed in action alone, to say nothing of wounded, and men dead from sickness and fatigue, while not a single Maroon was certainly known to have been killed. The situation was becoming serious: the negroes had begun to join the Maroons; the French might come at any moment; and then there would be every likelihood of a general revolt of the blacks against the whites, such as had already taken place in the Windward Islands. Walpole soon altered the whole plan of operations. He began by redistributing his posts, so as to command the mouths of the cockpits, employing negroes to clear away the jungle from the approaches and from the heights above them. He then set to work to train some of his men in the tactics of Maroon warfare, the essence of which was that men should work together in pairs or groups, one man taking charge of another's arms when he required both hands for climbing, and that above all they should take advantage of cover. Walpole had three infantry regiments with him; but the men that he chose for this work were the 17th Light Dragoons, and he did not regret his choice. So the two troops of the Seventeenth were dismounted and turned into mountaineer marksmen.

1795.

Colonel Walpole soon put his men into good heart by playing off the Maroon trick of ambuscades against themselves; for he lay in wait for one of their foraging parties, cut it off, and destroyed it to a man. A week later he sent a party of the Seventeenth along the right crest of the main cockpit in order to try and discover some fresh entrance into it. The party soon encountered the Maroons and became hotly engaged. The whole force of the Seventeenth numbered but forty men, of whom a fourth had been left in reserve under the command of a sergeant. Unfortunately, when called up in support, this sergeant led his handful of men straight into the mouth of the cockpit, where, of course, there was a bullet ready for every one of them. The main body, however, kept together, and was brought off in good order when compelled to retire by want of ammunition. Of the forty men one sergeant and three men were killed, and nine men wounded—a pretty heavy loss. None the less the Maroons were considerably dismayed by this bold attack, for hitherto they had been accustomed to lie hidden while the white men poured harmless volleys into the unresisting mountains. Still more dismayed were they when Walpole, having cleared the heights of jungle, managed by hook or by crook to get a howitzer in position and began to drop shells into the cockpit. In a very short time the Maroons were driven out of this favourite position, and compelled to withdraw to the adjoining cockpit. This was a serious matter for them, for the abandoned cockpit contained a spring of water. Walpole at once followed them up with the howitzer and drove them out of their second retreat. The Maroons then withdrew to a stupendous

1795. height so as to be out of reach of the shells. But a young cornet of the Seventeenth, Oswald Werge by name, saw one of the Maroon women leave the height to draw water, followed her unseen, and thus discovered the one path that led to the Maroon position. By this path the Seventeenth advanced, and again drove out the Maroons, who now retired down a very steep precipice into a third cockpit, where there was a spring of water. The Seventeenth occupied the abandoned height, and a detachment of the 62nd Foot under Colonel Hull marched into the virgin fortress of Cudjoe. They were the first white men who had ever penetrated into it, but they could never have entered it if the Seventeenth had not cleared the way.

What time was occupied by these operations, and with what loss to the Seventeenth, I have unfortunately been unable exactly to determine. There seems to have been a critical action on the 15th December, to which General Walpole makes allusion, but whereof no account can be found. All that is known is that thirty men of the Seventeenth, together with ten of another regiment (probably the 62nd) were posted so as to intercept the Maroons in one of Walpole's concerted movements, the whole detachment being under the command of a subaltern, who was not of the Seventeenth. The Maroons, however, managed to surprise this party, and shot down a certain number, including the officer, who, being disabled by his wound, made over the command to Sergeant-Major Stephenson of the Seventeenth. Stephenson was quite equal to the occasion. Far from being dismayed, he rallied his men and made a counter attack on the Maroons with a vigour that astonished them. Such conduct would have been creditable at any time, but it becomes particularly conspicuous when we think of the scare that had been created in Jamaica by the reputation and first successes of the Maroons. Stephenson was offered a commission in the infantry for his gallantry on this occasion, but stuck to his own regiment, in the hope of gaining a commission in the Seventeenth.

18th Dec. Three days after, Colonel Hull, still following up the Maroons with his little force of the Seventeenth and 62nd, fell in with them strongly posted on a precipitous hillside. The British halted on the acclivity over against them; and both sides opened a heavy fire. After about a dozen of the Maroons had fallen they ceased firing and began to blow their horns, as if desirous of seeking a parley. Thereupon the English fire was checked, and the Maroons were then told that the Colonel would grant them peace. For a long time they refused to believe it until Mr. Oswald Werge, of the Seventeenth, coolly threw down his arms, scrambled down to the valley below, and invited the Maroons to come and shake hands. It was an act of uncommon courage, for both sides, true to Maroon tactics, kept themselves carefully under cover; and therefore the first man to show himself, however pacific his intention, stood a good chance of being shot down. Werge's coolness, however, saved him. The Maroons took courage. One of them came down and shook hands with him, and presently exchanged hats with him, which was the Maroon symbol of perfect friendship. Thereupon it was agreed that hostilities should cease, and that Colonel Walpole should be sent for; and it was stipulated that neither British nor Maroons should advance until his arrival. Still neither force trusted the other; and, accordingly, the two tiny armies lay on their arms, weary, and worn and thirsty, to glare at each other through the livelong night. In the valley between them was a well; but in order that neither force should take an unfair advantage, it was agreed that British and Maroons alike should post two sentries over it. At length, however, the Maroons, unable longer to endure the agony of thirst, begged that the British sentries might be withdrawn while they drank, and engaged to withdraw their own in turn that the British too might drink. So both sides came down to the well and drank; and then the guard was posted again, and the rest returned to their arms. It must have been a strange scene, this of the rival sentries over the spring in that savage rocky glen—on the one side the wild negro of the mountain, his splendid athletic form barely concealed by a few foul rags, on the other the trooper

1795.

18th Dec.

1795. of the Seventeenth, bronzed, and lean, and haggard after months of harassing work, with his blue jacket faded, his white facings weeks soiled, his white breeches and Hessian boots sadly the worse for wear; but always erect and alert, and proud in the consciousness that he had beaten the dreaded Maroons on their own ground. There must have been good discipline in these sixty-four men of the Seventeenth and the fifty of the 62nd, seeing that with all the burden of a tropical climate on their backs they had outstayed the native mountaineers in the deliberate endurance of thirst within sight of water.

This action ended the war. The Maroons surrendered to Walpole, and submitted to beg His Majesty's pardon on their knees, while Walpole on his side promised that they should not be sent out of the island. This promise was violated by the Jamaica Government, whereat Walpole was so disgusted that he not only refused a sword of honour from the Jamaica Parliament, but resigned his commission. Thus the Seventeenth never had a chance of fighting under this gallant officer again. When he took charge of the operations the Jamaica Government was in such despair of quelling the Maroons that it actually imported a hundred bloodhounds from Cuba to hunt them down. When the hounds arrived the war was virtually over; and Walpole, in a letter to Lord Balcarres, has recorded to whom the credit was due:—

I must not omit to mention to your Lordship that it is to the impression made by the undaunted bravery of the 17th Light Dragoons, who were more particularly engaged on the 15th December, that we owe the submission of the rebels. The Maroons speak of them with astonishment. Mr. Werge was particularly signalised with the advanced guard, and the sergeant-major of that regiment is strongly recommended for his spirit and activity by the Commanding Officer, Mr. Edwards, who is in every way deserving of your Lordship's opinion.

CHAPTER VIII

GRENADA AND ST. DOMINGO, 1796

WHILE these two troops of the Seventeenth were making a name for the regiment in Jamaica, the remainder were very differently engaged. On the 6th August four troops embarked at Cork, 189 men being present and 194 absent in Jamaica and elsewhere, and sailed to Portsmouth, where they joined the cavalry camp at Netley, under Lord Cathcart. On the 21st September (according to the official record) they embarked for St. Domingo. From that date, if it be correct, it is extremely difficult to trace them. They formed part of the great expedition for the reconquest of the West Indies beyond all doubt; but that expedition did not sail until November, when the huge fleet of transports, under the convoy of Admiral Christian's squadron, was one of the most wonderful sights ever seen by Englishmen. The ships were not clear of the Channel before they were dispersed, many of them being lost, with appalling loss of life, by a storm. The fleet, all that was left of it, sailed again on the 9th December, and was again met by a storm, greatly damaged, and compelled to return to Spithead on the 30th. On the 26th December 100 transports were missing, of which no one knew whether they were afloat or gone to the bottom. It was not until the following March that Sir Ralph Abercromby, the Commander-in-Chief of the expedition, after having been a third time driven back to England by gales in February, contrived finally to reach Barbados, the headquarters of the British forces in the West Indies.

1795.

The Seventeenth, or at any rate some of them, appear to have

1795. reached the West Indies earlier than this. Two troops were employed, we are told, as marines on board H.M.S. *Hermione*, the ill-fated ship which in 1797 was the scene of one of the most disgraceful mutinies in the history of the British navy. Fortunately the Seventeenth had no share in the massacre of officers and delivery of the ship to the Spaniards, which make the name of the *Hermione* a byword. The two troops were landed at Martinique; but in order to understand why they were needed there it is necessary to glance at the history of the West Indies during the year 1795.

It has already been said that Mr. Pitt made early attack on the French Antilles. In addition to the expedition to St. Domingo, he in 1794 sent General Grey and Admiral Jervis to reduce the French islands of Martinique and Guadeloupe, which object they successfully accomplished. The adjacent islands of Grenada and St. Vincent had already been surrendered to us by France in previous wars, and were known as the French Ceded Islands. In 1795, however, the French contrived to stir up revolt against the English in the whole of these islands; and as in those days the French Revolutionists stuck at nothing, they did not hesitate to rouse the whole negro population, free and slave, against the British and ally themselves with it. The result was a quasi-civil war of the most barbarous kind—in fact, a turning loose of all the worst characters in the West Indies on the track of massacre and plunder. The garrisons of the British islands were so weak that in some cases, as in St. Lucia, they were overpowered and in others pressed to extremity. Grenada being the island wherewith the Seventeenth was engaged, it is necessary to glance at the course of the revolt therein.

Grenada, like most of the West Indian Islands, is simply a rugged, confused mass of volcanic hills, rising at their highest to three thousand feet. For the most part it is covered with jungle, but in the valleys and on the less precipitous ground the soil is fertile, and grows fine crops of sugar-canes and cacao. In shape the island is elliptical: it measures at its longest, from north to south, about twenty miles; at its broadest, from east to west, about ten miles. There are two little ports, St. Andrews and Grenville, on the windward or east side; another at the north point, Sauteurs; and two more on the leeward or western side, Charlottetown and St. George's, the capital. The garrison in 1795 consisted of 150 men of the 58th Foot, quartered in the barracks at St. George's, and in the old fort, called Fort George, which still commands the entrance to the harbour.

1795.

It was on the 2nd March 1795 that the revolt broke out in Grenada. None of the English had the least idea that it was coming. The Governor himself had gone away on a trip to the leeward side of the island, unconscious of any mischief. Before the morning of the 3rd of March had dawned the negroes had massacred the whites at Grenville Bay to windward, captured those at Charlottetown to leeward, and held forty-two of them, including the unlucky Governor, as prisoners in their hands. The civilian next in rank to the Governor at once took command of the island, sent to Martinique, Barbados, and Trinidad for assistance, and called out the local militia. This done he sent the 150 men of the 58th, together with the militia, to attack the insurgent post at Charlottetown. But when it came to the point the militia was not to be found—every man had fled on board the coasting vessels. The insurgents' position being very strong, the 58th could not attack it, and were compelled to return to St. George's.

12th Mar.

On the 12th March General Lindsay arrived from St. Lucia (which as yet was still quiet) with 150 men of the 9th and 68th Foot, and on the 17th attacked the insurgents, who forthwith retired to an impregnable position. Then the tropical rain came down and put a stop to all further operations. There are not many roads in Grenada now, and there were still fewer then —mere narrow, cobble-paved tracks, hardly wide enough for any wheeled vehicle. In fact these West Indies are miserable places to fight in, as this poor handful of British soldiers now discovered. Soaked with rain, exhausted by the stifling heat, and broken down by fever, the men had to tramp back as best they

1795. could. General Lindsay in the delirium of fever committed suicide, and his successor saw that without a stronger force it was useless to attack the rebels. Meanwhile the head of the insurgents, a ruffianly mulatto named Fédon, issued a proclamation threatening death to all who helped the English, and announcing openly that he would retaliate for any measures of repression by slaughtering his prisoners. As a natural consequence the negroes flocked to his standard in thousands, and laid the whole island waste.

1st April. On the 1st of April their arrived a weak reinforcement of the 25th and 29th Foot, probably about 400 men, from Barbados. With these and a few blue-jackets Brigadier Campbell attacked the insurgent stronghold on the 8th, but was repulsed. The rebel position was of extraordinary strength, well chosen, well fortified by abattis and other obstacles, and strongly manned. The British troops did all that men could do, with everything—numbers, climate, and tropical rain—against them; but they were compelled to retreat with the loss of 100 killed and wounded. Fédon then brought out his prisoners and cut the throat of every one.

Then, as usual, together with the rains came the yellow fever. The British troops suffered frightfully. "The 25th and 29th begin to fall down fast," says the General in a letter of 11th May. "Twenty died last week and six were carried off yesterday." So things went from bad to worse. No reinforcements could be obtained from the other islands, for one and all (excepting Barbados) were in a worse position than Grenada. St. Lucia had been evacuated; St. Vincent, after desperate fighting, was at the last gasp. In fact it seemed as if the West Indies were lost to England. By December the insurgent force in Grenada amounted to 10,000 men, well armed, furnished with artillery, and led by trained white French officers. The British troops, outnumbered on every side, were compelled to abandon the ports which they had tried to hold on the coast, and retire to St. George's. The rebels, or brigands as they were called, threatened to attack them even there. Nothing but the capture of the capital was wanting to give them absolute possession of the whole island. 1795.

But at last the tide began to turn. The long-awaited reinforcements from England had arrived at Barbados, and the relief of Grenada was at hand. On the 4th March 588 men from the 10th, 25th, and 88th Foot, under Brigadier Mackenzie, arrived at St. George's. They had lost 45 men in the course of a two days' passage; but their arrival was timely, for it compelled the insurgents to retire from before the capital. A week later further reinforcements from the 3rd, 8th, and 63rd Foot and the Seventeenth Light Dragoons landed at Sauteurs, at the extreme north point of the island. What were the numbers of the Seventeenth I have not been able to ascertain. One account says two troops, and I am inclined to think that this is correct. Whence these troops came, whether from England or Martinique, it is impossible to say. On the 24th March, pursuant to the designs of Brigadier Campbell, the forces at Sauteurs, 700 men in all, and those from St. George's, converged—the former by land, the latter by sea—upon the new position which the rebels had entrenched at Port Royal or Grenville. The troops, having been landed, worked during the night at the construction of a three-gun battery, and opened fire at daybreak next morning. But before attacking the main position on the principal heights, it was necessary first to clear some secondary heights adjoining them. For this duty the detachment of the 88th was detailed; but such was the difficulty of the ground that it was two hours before the 88th could even get near the enemy, and when they reached them it was only to be driven back. With great reluctance Campbell, who had made his dispositions not only to drive the rebels out, but to cut them off on every side, was compelled to bring up the 8th Foot to support their attack. Just at that moment a few of the rebels sneaked round to the rear of the British and set fire to the stores on the beach; and the conflagration was hardly extinguished when two French schooners anchored in the bay and began to land troops under cover of their artillery fire. Campbell saw that no time 1796. 25th Mar.

1796. was to be lost. Under a heavy cross fire from the rebel batteries ashore, and the guns of the schooners afloat, the Seventeenth charged down the beach and swept it clean, cutting down every soul. They then rallied and took post under cover of a hill. Meanwhile Campbell, quickly concentrating his infantry, led them straight to the assault, and, not without a severe struggle, carried the entrenchments by storm. The insurgents fled in all directions, but they did not get off scot free ; for, as they emerged upon the low ground, the Seventeenth swooped upon them and did great execution. Three hundred brigands, mostly *sans-culottes* from Guadeloupe, are said to have met their fate at the hands of the regiment that day. No prisoners were taken : it was not a time for taking prisoners ; and the survivors of the pursuit took refuge in their original stronghold opposite Charlottetown. The total British loss was 12 officers and 135 men killed and wounded. The Seventeenth lost but 4 men wounded, one horse killed, and two horses wounded; but the detachment, together with its commander Captain John Black, was highly commended both in orders and despatches for its behaviour in the action.

After this engagement nothing more was done for a time, owing to the general confusion caused by the revolt. The Seventeenth was moved to St. George's and quartered in Government House, much to the disgust of the new Governor, who arrived in April and wanted the house to himself. Meanwhile the main expedition under Sir Ralph Abercromby had at 17th Mar. last arrived from England and was concentrating at Barbados. He turned his attention first to St. Lucia, which was recaptured on the 24th May, and then to St. Vincent, which was finally 19th June. relieved on the 10th June. A few days later he sent a force to Grenada, which landed at Charlottetown and advanced upon Morne Quaqua, the great rebel stronghold, from the west, while a second column moved against it from the east. This Morne Quaqua was a remarkable position. The rebel camp was on a height at a considerable elevation, and above it rose a rocky precipice accessible only by a narrow path, which path, together with the lower ground beneath it, was commanded by a field-gun and several swivels and wall-pieces. Above this rose another bluff with another gun in position, and finally above this again, at the head of a very steep ascent, came the summit. Felled trees and abattis made good any points that nature might have left unstrengthened. Nevertheless, the French commandant, when he saw the advance of the British columns, lost heart and surrendered. Fédon and the desperate faction thereupon led out their English prisoners, some twenty in number, stripped them, bound them, and murdered them. They then fled to the jungle, where they were hunted down by the troops and hanged in twos and threes. Fédon alone, most unfortunately, was never caught.

1796.

So ended the relief of Grenada, wherein the Seventeenth took decidedly a leading part. How long the detachment remained in the island it is impossible to discover, but probably not for very long; for by August, so far as can be gathered from scattered notices, five troops of the regiment were at St. Domingo and three at Jamaica. It is to these three latter that a muster-roll taken in December 1796 most probably refers,—a ghastly document wherein, unfortunately, the place of muster is not mentioned. It shows that between 25th June and 24th December 1796, of—

12 sergeants 7 died,
116 privates 76 died,
2 trumpeters both died.

Thirty-seven men out of 130 died in a single week, and but forty-five were left alive when the muster was taken. Captain John Black, who had done so well in Grenada, was dead by July; one of the Lieutenant-Colonels, George Hardy, had died a month before him. Such was yellow fever in the West Indies a hundred years ago.

Of the services of the regiment in St. Domingo it has been extremely difficult to gather any information, owing to the absence of all St. Domingo despatches from the Record Office. It would appear, however, that the Seventeenth was quartered at

1796. Jeremie under the orders of General Bowyer. The French, under the command of the coloured man Rigaud, were very active, in the spring of 1796, in attacking the various scattered posts occupied by the British on the south-eastern promontory of St. Domingo, round about Jeremie. In August, General Bowyer being 8th Aug. apprehensive of further attack on these posts, sent Captain Whitby with two subalterns and sixty rank and file of the Seventeenth, dismounted, eastward to Caymites, *en route* for the two posts named Fort Raimond and Du Centre. At this latter place they 10th Aug. arrived on the 10th. Whitby had hardly time to send a small detachment of the 13th Light Dragoons to Raimond, when that post was attacked by the French, who were repulsed with severe loss. Whitby then reinforced Raimond still further by a detachment of twenty men of the Seventeenth under Lieutenant Gilman, who took post in the block-house. On the 12th the enemy were still before the block-house, keeping up a heavy though not very effective fire, when Gilman at last grew tired of it, sallied out with his twenty men of the Seventeenth and a few Colonial irregulars, and drove them off into the jungle. The French left a small field-gun behind them, and sixty-three dead on the field, sixteen of whom were whites. Many more dead and wounded were found dead in the jungle afterwards. "I am happy to say," wrote General Bowyer, "that in this gallant affair the Seventeenth had only two privates wounded. Lieutenant Gilman's[1] cool conduct and intrepidity on this occasion seem to me so praiseworthy that I should not do justice to my own feelings if I did not recommend him for promotion."

Simultaneously Bowyer was under the necessity of raising the siege of Irois, another post, which Rigaud had besieged for eighteen days with 4000 men. Then hearing that the French had taken up a strong position on a mountain called Morne Gautier, to cut off communication between Irois and Jeremie, he resolved to attack it. He therefore marched in three columns

[1] This officer was not of the Seventeenth.

at daybreak on the 16th August, and opened fire at long range. 1796. Seeing that the men of the Seventeenth, who formed part of his force, were falling fast, he determined to carry the position by assault, and had formed the Seventeenth for the purpose, when he was disabled by a bullet which struck him in the left breast. None the less the attack was made; and though the British were driven back the French retreated in the night, and Irois was saved. In the course of these operations the Seventeenth lost about thirty men killed and wounded, seven having been killed and fifteen wounded in the attack on Morne Gautier alone. As only half the regiment was in St. Domingo, and that half terribly reduced by sickness, these losses cannot but represent at least a third, if not more, of the numbers engaged.

With this the record of the Seventeenth in St. Domingo comes to an end. What further work it may have done is buried in the lost despatches and the lost regimental papers. There is a complete 1797. muster-roll of the regiment dated Port Royal, 4th March 1797, showing that 126 men died in the course of the year 1796; but whether the regiment was moved thither from St. Domingo before its return home, or whether it sailed home direct, must remain uncertain. In any case it left the West Indies, and arrived in England in August 1797. The bad luck at sea which had marked the departure from England attended the passage home. The headquarter ship, the *Caledonia*, foundered at sea, and though the men were saved the baggage and regimental books were lost. Hence the scantiness of information respecting the first forty years of the life of the regiment.

CHAPTER IX

1797-1807

OSTEND—LA PLATA

1797. ON landing in England the Seventeenth was distributed into quarters at Nottingham, Leicester, Trowbridge, Bath, and Bristol. The regiment was reduced to a mere skeleton. Four hundred recruits and a draft from the 18th Light Dragoons, however, soon filled up the gaps and restored it to its strength. All ranks had something new to learn. In 1796 a new drill-book, far more ambitious than any that had yet appeared, was provided for the cavalry; and for the first time (so far as I have been able to discover) a properly authorised system of sword exercise. The drill shows little that is new, except that the system of telling off by threes now came into general use, and with it the practice of executing all movements to the rear by means of "Threes about." The interval of "six inches from knee to knee" in the ranks also makes its appearance as the normal formation. A further change is the reversion to the old practice of posting troop leaders on the flanks of troops, dressing with the men, and covered by a corporal in the rear rank.

As regards sword exercise we must content ourselves with observing that we encounter for the first time the once famous "six cuts." The recruit was posted in front of a wall on which was drawn a circle; and he was then taught that each of the six cuts required of him should intersect at the centre of the circle, and divide it into six equal segments. I do not mean that the unhappy man was tortured by any such abstruse terms as these, but that this was the principle on which the six cuts were based. In addition, there was a seventh cut, directed vertically, so to speak, from heaven to earth, and called by the high-sounding name of St. George. These seven cuts are still familiar to hundreds of living men. The whole of the sword exercise was comprehended in no fewer than six divisions, each containing from seven to ten words of command, and must therefore have consumed considerable time. It may be remarked that, when cutting the sword exercise on foot, the men were not required to extend their legs as at present, though they kept the bridle hand in the bridle position. The swords themselves were perhaps the most defective part of the whole concern, and caused great complaint among the Light Dragoons in the Peninsula. The pattern was bad, and the material was bad; and common sense was so absolutely ignored in the design that the hilt was not even provided with a guard. Before quitting the question of drill, it is well to remind readers that dismounted drill still occupies a prominent place in the training of the Light Dragoons; and the words "Form battalion" and "Fix bayonets" are still in full use. 1797.

In 1798 the regiment was moved to Canterbury, where it made the acquaintance of a naval officer who was destined to exert some influence on a part of its career. This was Captain, afterwards Sir Home, Popham. Just then he was full of a scheme for blowing up the lock-gates of the Bruges Canal, which lock-gates were situated at Saas, a village just a mile from the entrance to Ostend harbour. The canal itself from Bruges to Saas was thirteen miles long, one hundred yards wide, and thirteen feet deep, and had recently been completed at a cost of five millions. For the invasion of England it was of great importance to the enemy; for any number of vessels could be fitted up therein and brought down to Ostend without risk of facing the British cruisers at sea. If an invasion were intended, Ostend was obviously the best port of embarkation for the invading army; and even if the project of a descent on England should prove 1798.

1798. to be no more than a scare, the destruction of the lock would at any rate spoil a seaport and stop all internal navigation from Holland to West Flanders.

April. So Captain Popham argued; and his arguments were held to be good. Accordingly the whole plan of operation was entrusted to him; and preparations for the little expedition went forward with the utmost secrecy all through the month of April. By the second week in May everything was ready, and on the 13th the troops were embarked at Margate on seven transports. The force consisted of four companies of the 1st Guards, the flank companies of the Coldstream Guards, 3rd Guards, 23rd, and 49th Foot; the 11th Foot, artillerymen with six guns, and, lastly, one sergeant and eight men of the 17th Light Dragoons, the only mounted men of the expedition. On 16th May. the morning of the 16th May the little fleet got a fair wind and sailed away, arriving, without further mishap than leaving the 1st Guards hopelessly astern, in Ostend at 1 A.M. on the 19th May. 19th. For a time everything went like clockwork. Sir Eyre Coote, who commanded the expedition, summoned the French commander at Ostend to surrender, as a feint, to make him believe the town was the object of attack. Then having received a high-flown reply, and seen all the French troops drawn into Ostend, he quietly landed his men on the opposite side of the river, and blew up the lock-gates with the greatest success. By 11 A.M. Coote was back on the beach and anxious to re-embark, having accomplished his object with the trifling loss of five men killed and wounded. But meanwhile a gale had sprung up, and the surf was so great that re-embarkation was impossible. After several futile attempts, in which boats were swamped and the men nearly drowned, Coote decided to entrench himself where he lay and wait for better weather.

20th May. At four o'clock next morning, when the wind and surf had considerably increased, the enemy was seen advancing in two columns, with far superior numbers, against Coote's position. Outnumbered and outflanked the British force fought for two hours against hopeless odds, until Coote was wounded while rallying the 11th Foot. Then General Burrard, the second in command, seeing the front broken and both flanks turned, was compelled to surrender. Of the 1100 men landed, 163 were killed and wounded, and the rest of course taken prisoners. Of the nine men of the Seventeenth, one was wounded. So exemplary had been their behaviour, we are told, that when, shortly after, they were exchanged and returned to the regiment, every man of them was promoted to be a non-commissioned officer, while the sergeant, William Brown, was given a commission, first in the waggon train and latterly in the regiment. As usual the non-commissioned officer of the Seventeenth, when in independent command, brings credit to his corps.

1798.

1799.

In this same year two squadrons of the regiment were ordered to Portsmouth to embark for Egypt, but, the order having been countermanded, the whole regiment joined a large cavalry camp then formed at Swinley. In the following year another camp of 30,000 men was formed on Bagshot Heath under the command of the Duke of York, of which the regiment again formed part. In September it was employed in suppressing riots which had arisen in consequence of the high price of provisions. While engaged in this service many men were badly knocked about, and Captain Werge, who had escaped without injury from such deadly marksmen as the Maroons, narrowly escaped death at the hands of his own countrymen, receiving a shot through his helmet. Two troops having been added to the establishment, the regiment paraded in its greatest recorded strength at Manchester in the following year—upwards of 1000 non-commissioned officers and men, and nearly 1000 horses, being present. Colonel Grey was the fortunate officer who held command, and we must hope that Major-General Oliver Delancey, the Colonel-in-Chief, who alone could remember the regiment before it went to the American War, went up to inspect so fine a corps. Unfortunately this magnificent strength did not last long. In May 1802, England and France, being both of them

1800.

1801.

1802.

1802. exhausted after nine years' fighting, agreed to the peace of Amiens. Thereupon, with the usual blindness, the army was reduced, and two troops of the Seventeenth were disbanded. Their horses were valued by a dealer at forty guineas apiece, a larger price in those days than in these, which shows that the regiment must have been superbly mounted.[1]

1803. Peace lasted for just fourteen months; and then in May 1803 England took the initiative and declared war against France. On the 1st of that month the Seventeenth embarked from Liverpool for Ireland. It met with its usual luck at sea on the passage, the transports being dispersed by a gale which drove them into various ports on the East Coast, and permitted but one immediately to reach its destination at Dublin. In 1804. the course of the following year the establishment was again augmented to ten troops, four of which joined the camp at the Curragh, where a large force was assembled under the command of Lord Cathcart. This Lord Cathcart, let us remember, was an officer of the Seventeenth during the American War; he is the same man who commanded the expedition against Copenhagen in 1807, when Sir Arthur Wellesley himself served under him. 1805. The following year is memorable for the formation of Napoleon's camp of invasion at Boulogne. Napoleon's hopes having been shattered by Nelson's victory at Trafalgar (12th October), he broke up the camp and marched away to the campaign of Ulm and Austerlitz. Previous to these two great disasters there had been some idea of a diversion to be made by an English army on the Continent; and in September the Seventeenth received orders to prepare for foreign service as part of this force. But Austerlitz effectually smothered this design. In December the regiment was moved back to England, and spent Christmas day on the passage, the first of four successive Christmas days that it was destined to celebrate on the sea.

1806. The year 1806 opened gloomily with the death of William

[1] This year 1802 also witnessed the introduction of the chevron on the sleeves of non-commissioned officers.

Pitt, the great man whose indomitable spirit had carried England through the first and worse half of the tremendous contest against France. The want of his guiding hand was soon to be badly felt. 1806.

The month of March brought a nearer occasion of mourning to the Seventeenth. On the 20th there died at the Plantation, Guisbrough, in Yorkshire, General John Hale, the father of the regiment. He had been promoted Major-General in 1772, Lieutenant-General in 1777, and General in 1793, and, it seems, had settled down to end his days among his wife's people. In his long life of seventy-eight years he had seen the rise of William Pitt, "the terrible cornet of horse," and the death of his son William Pitt, "the pilot who weathered the storm." He left behind him seventeen children and the Seventeenth Light Dragoons.

Just about this time unfavourable reports of the regiment found their way to headquarters, insomuch that a general was sent down to Northampton to inspect it. Rather to his surprise this officer found that, so far from being unfit for active service, the regiment was the best in the matter of men and horses, drill and equipment, that he had seen. He reported accordingly to headquarters, with results that were speedily apparent. April.

In September, the regiment being then distributed in quarters at Brighton, Hastings, Romney, Rye, and other points on the south-east coast, there arrived suddenly one night an express message ordering the Seventeenth to prepare forthwith for foreign service. Its route, it was added, would be sent down immediately. On the 27th September the regiment marched to Portsea and Southampton, and having detached two troops to Chichester as a depôt, gave up its horses and embarked on the 5th October at Spithead, bound for South America. It must now be explained where and why it was wanted. 27th Sept.

On the 4th January 1806, just when the Seventeenth was disembarking in England from Dublin, there arrived off the Cape

1806. of Good Hope 4000 British troops under Sir David Baird, convoyed by a squadron under Commodore Sir Home Popham. The troops were landed; and in less than three weeks the Cape Colony had passed from the Dutch into the hands of the English for ever. Before he sailed, Sir Home Popham, always a busy man, had become greatly bitten with the idea of an attack on the Spanish possessions in Central and South America, that is to say, on any part of Central and South America except Brazil, which was a Portuguese Colony. He had held many conversations with one General Miranda, a native of Venezuela, who was at the head of a revolutionary movement against the dominion of Spain in South America, and had promised that if the British would send a force thither the whole population would rise and fight at their side against Spain. It was the old story which had taken the English to the Carolinas in 1781, and to St. Domingo in 1793, with most disastrous results. But Popham, forgetting these two lessons, continually urged upon the English Government the project of an attack on South America, and even drew up a complete plan of operations for descent on the continent from the Atlantic and Pacific sides simultaneously.

The date of this plan is October 1804. The memorandum had been before the British Government for more than a year, and had received little or no notice. At three months' distance from England, with men and ships to his hand, and no one in command over him, Popham persuaded Baird to let him have Brigadier-General Beresford (afterwards well known in the Peninsular War as Marshal Beresford) and 900 men; and with 14th April. these and his squadron he sailed away for Rio de la Plata, to take Buenos Ayres on his own responsibility. At first everything went well. The force, strengthened by 200 more men picked up at St. Helena, duly arrived in the Plata, and disembarked on the 25th June at a point ten miles below Buenos Ayres. From thence, in spite of Spanish troops in greatly superior numbers that were drawn up to oppose him, Beresford marched practically unchecked and unhindered into the city, and on the following day received its surrender. 1806. 26th June.

For seven weeks Beresford held Buenos Ayres, the people swearing allegiance to King George, and doing everything in the way of promises that was asked of them,—all of which did not prevent them from rising *en masse*, when their preparations were complete, and attacking Beresford with unmistakable fury. With but 1300 men against 13,000, Beresford fought for three hours and inflicted heavy loss on the enemy, but having lost 12 officers and 150 men, he was at length compelled to surrender. The Spaniards agreed to his proposals that he and his army should be shipped off to England forthwith; and there it might have been supposed that the whole matter would have ended. But it was not to be. The Spaniards most treacherously violated the treaty, and carried off Beresford and the whole of his army into the back country as prisoners. 12th Aug.

On the first capture of Buenos Ayres Popham had, of course, sent despatches home to report his success. The Government, however, was, for various reasons, much annoyed and embarrassed at Popham's escapade, and responded by ordering him to England and trying him by court-martial. Still the nation at large was so delighted at the exploit that the Government, after much hesitation, was forced to send out reinforcements under Sir Samuel Auchmuty. Auchmuty's instructions bade him simply make good Beresford's losses and await further reinforcements, failing the arrival of which he was to proceed with his troops to the Cape. At one moment in August the whole expedition was countermanded; but finally the Government made up its mind and decided, on 22nd September, to despatch it. This vacillation accounts for the very short and sudden warning received by the Seventeenth. The whole force under Auchmuty's command numbered 3000 men, viz. the Seventeenth, 700 strong; the 87th and 40th regiments of Foot; three companies of the 95th (now the Rifle Brigade), and 170 Artillery. The transports finally sailed from Falmouth on the 9th October, the British Govern-

1806. ment being still in ignorance of the loss of Buenos Ayres and of the capture of Beresford's army.

The haste in the equipment of the expedition soon showed itself in various ways. The transports were such miserable sailers that, long before they reached their destination, they ran short of water, and were obliged to put in at Rio Janeiro. There Auchmuty heard of Beresford's disaster, and further of the arrival of a small reinforcement of the 47th and 38th Foot, which had been sent from the Cape to the Plata, and had taken up a position at Maldonado, a town standing at the entrance to the river on the north side. Not knowing what to do, Auchmuty victualled his ships for four months and started off again for Maldonado, where

1807. 5th Jan. he arrived at last, after a passage of 147 weary days, on the 5th January.

Finding that Maldonado was an untenable position, Auchmuty

13th Jan. evacuated it a week later and sailed up the river. The retention of Beresford's army was an act of treachery which called for reprisals, and these he resolved to take by attacking Monte Video, which stands on the north bank of the river, on the opposite side to Buenos Ayres, and some one hundred and twenty miles below it. On the 16th he landed in a small bay to west of Caretas Rocks, nine miles from Monte Video, the enemy watching the disembarkation in great force, but not daring to oppose it. Three days later Auchmuty began his advance upon Monte Video in two columns, the right column being made up of the Seventeenth, two troops of the 20th, and as many of the 21st Light Dragoons, all of them dismounted, under Brigadier-General Lumley. The Seventeenth had previously exchanged their carbines for Spanish muskets, which had been obtained at Rio Janeiro. This right column was early attacked by the enemy and threatened by 4000 Spanish cavalry, which occupied two heights in the front and right of Auchmuty's advance. The attack, however, was soon repulsed by the dismounted cavalry and the light companies of the infantry; and the enemy retired, allowing the British advanced posts to occupy the suburbs of

Monte Video on the same evening. Auchmuty himself had his horse shot under him while directing this column, and remounted himself on Colonel Evan Lloyd's charger. 1807.

Next day the enemy took the initiative, sallying forth against Auchmuty's force with 6000 men and several guns. This time they attacked the British left and left flank with cavalry, using their infantry to keep the dismounted cavalry in check. After driving in the picquets the Spanish infantry column was repulsed with great slaughter, and the cavalry then retired. The enemy's loss in this action was reckoned at 1500. The English loss between the 16th and 20th was 18 killed and 119 wounded of all ranks. 20th Jan.

Arrived before the town, Auchmuty discovered that the defences of Monte Video were not "weak," as Popham had described them in his memorandum, but, to use Auchmuty's own word, "respectable," mounting 160 guns. Moreover the Spaniards, through possession of a fortified island, kept command of the sea, and were able to cannonade the British advance from their gun-boats. Nevertheless, Auchmuty was fully decided that he would take Monte Video somehow. While he was making up his mind how to do it the enemy appeared on his rear, but was repulsed after a sharp skirmish, in which the Seventeenth lost a few men. After a few days' construction of batteries and other preparations, Auchmuty saw that if Monte Video was to be taken it must be stormed, and accordingly made his dispositions for an assault at daybreak on the 3rd February. Naturally he chose infantry regiments for infantry work, and left the Seventeenth, together with the rest of the cavalry, the 47th Foot, one company of the 71st, and 700 marines to protect the rear and cover the attack, under the command of General Lumley. The storming force did its work magnificently, and in a few hours Monte Video was in Auchmuty's hands, though at the cost of 27 officers and 370 men killed and wounded. 22nd Jan. 3rd Feb.

Horses being cheap, some of the Seventeenth were now mounted, doubtless a very welcome change from the drudgery of the infantry work during the siege of Monte Video; though even

1807. when employed on foot the regiment earned the personal thanks of the General. The Seventeenth had shown that it could beat the infantry at its own work in Jamaica eleven years before. But the native South American horses, as Auchmuty himself says, were not strong enough to carry the equipment of the British dragoons. The native irregular horsemen, armed with muskets and swords, pursued a method of warfare of the most harassing kind. They would ride up in twos or threes, dismount, fire over their horses' backs, mount again, and gallop off before the British had a chance of catching them. And these men were not soldiers ; they were the ordinary members of the population, not friendly as Popham had hoped, but inveterately hostile to the European invaders. In fact the British on the Plata found exactly the same elements opposed to them in New Spain as Napoleon was to find, a few months later, in the old Spain which is known to us March. as the Peninsula. Owing to the difficulty of obtaining forage, the mounted men of the Seventeenth, some 220 in number, were sent up the country forty or fifty miles from Monte Video to Lanelones and St. Joseph, while the remainder of the regiment was quartered in and about Monte Video.

Meanwhile, since the departure of General Auchmuty, the British Government had committed itself to the project of a general attack on Spanish South America. Sir Arthur Wellesley himself was called upon to give advice respecting it. Finally, on the 30th October General Craufurd (the famous Craufurd of the Light Division) was ordered off with 4000 men, with instructions to take Lima and Valparaiso on the Pacific coast, and to open communications with Beresford across the continent when Valparaiso was in his hands. Craufurd sailed on the 13th December 1806, arrived at Porto Praya on the 11th January 1807, waited for several weeks there in vain for the admiral who was to go with him, and at last in despair sailed for the Cape, where he arrived on the 20th March. There he found orders to join Auchmuty at Buenos Ayres, and accordingly sailed thither on the 5th April. The confusion caused by the efforts of the

British Government to manage a campaign at from three to six months' distance from England, can be appreciated only by those who have read the original despatches.

1807.

In February there arrived in the Plata a reinforcement consisting of the 9th Light Dragoons, a fact worth noting, inasmuch as this is the only occasion on which this great regiment, the first of the Lancer regiments, has fought side by side with the Seventeenth. The 16th and Seventeenth fought together in their youth in America. Thus after unspeakable confusion a large British force was at last in process of concentration on the Plata. And now the Government in an evil hour decided to put another commander over the heads of Craufurd and Auchmuty, and chose for the purpose General John Whitelocke. He arrived on the 10th May, and found that Auchmuty had already seized the town of Colonia, immediately opposite to Buenos Ayres, so as to make the passage across the river as short as possible. A month later Craufurd arrived, and next day the Seventeenth and the artillery were embarked at Monte Video, while the rest of the army moved up to Colonia to embark there. Devoutly thankful the Seventeenth must have been to get to serious business again. Forage was terribly scarce for the horses, and flour hardly less scarce for the men, though bullocks could be bought for a dollar a head.

15th June.

The passage up the river was delayed by contrary winds, but at last the hundred miles were traversed, and the troops landed at Ensenada, thirty miles below Buenos Ayres. The moment the army was disembarked it was surrounded by a cloud of Spanish light cavalry hovering about just out of musket range. Here was the opportunity for using the Seventeenth; but it was not employed. Two of the four mounted troops, each of forty men, were ordered to give up their horses to the commissariat. But when the pack-saddles were put on them the horses broke loose, and were from that moment useless. Thirty more mounted men were detailed to look after the landing of provisions, of whom ten were used as orderlies to carry despatches. Twelve more were attached to one of the infantry brigades; and the remainder, forty-eight all told,

28th June to 5th July.

1807. accompanied General Whitelocke, principally, no doubt, as his escort. The natural consequence was that the army could hardly advance at all. One staff officer was taken prisoner by the enemy's light cavalry while carrying orders between two brigades, and another was stabbed within three hundred yards of the flank of the British line, all for want of a little cavalry which, with unspeakable folly, had been dismounted just when it was most sorely needed to encounter the enemy's horse.

On the 29th June the advance began, across a very difficult country, much intersected by ditches and swamps, the dismounted men of the Seventeenth forming the rear-guard. The army was like to have been starved on this short march, but eventually it reached Buenos Ayres, after brushing aside some slight opposition from the Spaniards on the 4th July. Part of the Seventeenth and 40th Foot were left behind at the village of Reduction on the way, to protect the artillery. Sixteen of them, mounted men, together with thirty dismounted men of the 9th, were engaged in repelling an attack on the rear of the British advance.

3rd July. On the 3rd July General Whitelocke managed to lose his army; but on the next day he found it again, and on the 5th July made his attack on the city. That is to say, that he sent 6000 men up fourteen different streets through three miles of a hostile town, with strict orders not to fire until they reached the far end. What is more, the 6000 men did it. Nearly every street was entrenched and defended with cannon; every house was strongly barricaded and a fortress in itself; from every roof came a shower not only of bullets but of stones, bricks, and tiles, and every description of missile. Nevertheless the men did fight their way to the other end of the town without firing a shot; but by the time they had reached their allotted positions 1000 of them were down, and 1500 more, Craufurd himself among them, had been overpowered and compelled to surrender. Nevertheless Auchmuty on the left held a strong position, to which many men had rallied, where he had captured 32 guns and 600 prisoners; and with him sixteen mounted men of the Seventeenth, together with some

5th July.

infantry, opened communication, through all the fire, from the reserve. On the extreme right the British also held a strong position, and thither also some mounted men of the Seventeenth made their way from Reduction, to keep in touch with the city. But all was to no purpose. Next day Whitelocke came to terms with the Spaniards, and agreed to withdraw every British soldier from the country.

1807.

So ended the ill-fated expedition to the Plata. Whitelocke was tried by court-martial on his return, and cashiered. The British in any case could hardly have kept a hold on the country; but Popham's error was no excuse for Whitelocke's mismanagement. This was the third time in fifty years in which the Seventeenth was sent on a fool's errand to a country where the population was expected to receive them with open arms, and met them in fact with loaded muskets. Carolina in 1781, St. Domingo in 1796, and the Plata in 1806, were all part of one great blunder; and for all three the Seventeenth suffered. It is not a soldier's business when sent on active service to inquire as to the wisdom or unwisdom of the statesmen who send him. He must simply obey orders, and do his duty. But it is hard when years of good and gallant service by a regiment are buried under the cloud of a statesman's blunder; and this has been the fate of the Seventeenth.

CHAPTER X

FIRST SOJOURN OF THE 17TH IN INDIA, 1808-1823—THE PINDARI WAR

1807. THE army evacuated the Plata in November. The Seventeenth was driven by stress of weather into Cork Harbour, and thus spent 1808. their second consecutive Christmas Day on shipboard. Leaving Cork early in January it sailed to Portsmouth, disembarked on the 17th, and joined the depôt troop at Chichester, where it remained for six weeks dismounted under orders for the East Indies. Every man who asked for a furlough within a hundred miles of London obtained it; and this was well, for there were not many of them that saw their homes again. Still, though the furlough was extended to the 20th February, every man, with the exception of one detained by sickness, was present at the expiration of the term. Moreover, though the men had money in their pockets, having arrears of pay due to them on their return, there was not a single case of misconduct at Chichester; and that meant a great deal in these hard-drinking days. The men had gone through much since they were last in England—147 days at sea in miserable transports, most of the time within the tropics; then a campaign with plenty of hardships and very little glory, wherein their horses were taken from them just when they could have been most useful; then a two months' passage home in bad weather, and the mortification of landing as part of an unsuccessful army, and unsuccessful through no fault of its own. Finally it was under orders to sail in six weeks to the East Indies, a very deadly quarter to Europeans in those days.

The Mayor and Corporation of Chichester could not understand how a regiment in such circumstances could spend £3000 in the town in six weeks without a single instance of misbehaviour, and went so far as to express their thanks to the Seventeenth for its exemplary conduct.

1808.

29th Feb.

A few days later the regiment embarked at Portsmouth, 800 strong, under the command of Major Cotton; Lieutenant-Colonel Evan Lloyd being detained to give evidence on General Whitelocke's court-martial. On the 1st of June it arrived at the Cape of Good Hope, where it found one of its old colonels, Major-General H. G. Grey, and was inspected by him. From the Cape the regiment sailed for Calcutta. As it was approaching the Hugli one of the transports, the *Hugh Inglis*, was set on fire by the carelessness of a petty officer, but the fire was extinguished without serious damage. Next day the three topmasts were carried away by a squall, and swept fourteen or fifteen men overboard with them, of whom, however, all but one were saved. The Seventeenth has gone through a good many adventures at sea between gales, founderings, fires, and service as marines.

4th June.

On the 25th August the regiment was disembarked at Calcutta, 790 men strong, and did garrison duty in Fort William until December; during which time Major Cotton, the regimental quartermaster, and sixty-two non-commissioned officers and men, fell sick and died—a melancholy opening to its first term of Indian service. In the following year it was placed on the Bombay establishment, and sailing from Calcutta arrived at Bombay on the 1st February. From thence it was moved up to its destined quarters at Surat on the Tapti River, some two hundred miles north of Bombay. Two galloping guns worked by its own men were added, as was usual, to the establishment; and by a concurrence of testimony the regiment was excellently mounted.

1809.

Early in 1810 the Seventeenth was employed on a rather curious service. At the end of 1809 there was a sudden rising of religious fanatics in Mandavi under the leadership of a man named Mean Abdul Rahman, who killed the vizier of

1809. Mandavi, and put the rajah to flight. The leader then sent a message to the English Resident, ordering him to accept Mohammedanism or fight. He added that he was come down to earth in the bodies of four great men, Adam, Jesus, Ahmad and Mean Abdul Rahman, and concluded with a request for three hundred rupees. Absurd as the matter sounds, it soon assumed a 1810. serious aspect. The news of the rising reached Surat on the 10th January, and the people at once flocked out from the city to join the new prophet. The Mohammedans in general began to assume a threatening attitude, and attacked the Hindoos with the cry of "Deen." In fact there were the elements of a troublesome disturbance, which in the judgment of the Resident required to be suppressed at once. Accordingly four troops of the Seventeenth, under Major Supple, and some infantry were called out and marched off to the village of Boodham, where the prophet and the most devoted of his followers were assembled. The Seventeenth outmarched the infantry, and came up with the fanatics at day-break on the morning of the 19th January on the plain outside the village. The fanatics were summoned to surrender and give up their leader; but they replied with shouts of defiance. A feint attack was then made to intimidate them; but they simply threw up clouds of dust in the horses' faces and dared the Seventeenth to the combat. Then the regiment attacked in earnest, and there ensued what the Resident called a "furious engagement." The fanatics were armed with spears and small hatchets attached to bamboo shafts, twelve or fourteen feet long, with which they could inflict severe wounds; and they fought like demons. If the Seventeenth had had lances in these days they might have made short work of them; but, as things were, the fighting lasted for some time. It was not until 200 of the fanatics lay dead on the field that the bulk of them dis-persed and fled to the village, where, still undefeated, they renewed the fight against the infantry and artillery. Finally the Seventeenth set fire to the village and put an end to the affair; and the leader of the fanatics, having been wounded in the first action, was captured by the infantry. Of the Seventeenth, one corporal and two privates were killed; all the officers, several privates and many of the horses were wounded. Lieutenant Adams' helmet was cut to pieces on his head. 1810.

In this same year a detachment of the Seventeenth, under Lieutenant Johnson, accompanied Brigadier-General Sir John Malcolm on his mission to Persia. On its return in December this detachment brought with it a letter from Sir John to the Colonel, in which the former went out of his way to express his high opinion not only of Mr. Johnson, but of the non-commissioned officers, Sergeant Willock and Corporals Carrigan and Batson, who were with him. It is remarkable to note that non-commissioned officers of the Seventeenth, employed with small detachments, have never failed from the first to command the admiration of all strange officers whom it has been their duty to serve. A curious memorial of this escort was found in the ruins of Persepolis by an officer of the regiment (Lieutenant Anstruther Thomson, now Captain Anstruther) while travelling in 1888. Scratched on one of the lions at the head of the main stairway are the death's head and cross-bones with the motto, and beneath it the name "Sergt. Robt. Willock"; and on the wall of Xerxes' house is cut the name of "Pte. M. Cloyne, 17 L. Ds. 1810."

Before we quit this year we must add two small extracts (copied from the *Calcutta Gazette*) from the Dress Regulations, which gives us a faint glimpse of the transition through which the British Army was passing:—

10*th October*.—Clubs and queues are abolished in all ranks from this date, and the hair is in future to be cut close to the neck. No powder is to be worn on duty.

This is the first beginning of the short hair, which now particularly distinguishes a soldier. Old as the queues were, the whole Army was delighted to be rid of them, though there were antique officers that regretted them to the end. At the beginning of the great war with France the War Office, which was decidedly

1810. negligent in the matter of feeding the troops in Flanders, never failed to send them shiploads of leathern queues.

8th November.—Scale epaulettes are to be worn exclusively by officers of cavalry.

No shoulders have seen more vicissitudes of adornment than those of the British officer.

1811. In December of the following year the regiment left Surat for new cantonments at Ruttapore, near Kaira, in the northern 1812. division of Guzerat. On the 1st of January following Lieutenant-Colonel Evan Lloyd was promoted to be Major-General, and retired from the command. He was the last of the officers then doing duty with the regiment who had served with it in the American War. His successor was the Hon. Lincoln Stanhope, who came from the 16th Lancers, and was blamed by his brother officers in that corps, not without justice, for preferring "an arduous campaign in Bond Street" to duty with his regiment in the Peninsula. None the less he did good service enough with the Seventeenth.

The year 1812 brought with it a further change in the clothing. The cord lacing and the innumerable buttons that had adorned the front of the jacket were abolished, and another jacket with broad, white facings, almost as wide as a plastron, was substituted in its stead. Simultaneously the old helmet disappeared and the felt shako took its place. The old white breeches and knee-boots were likewise swept away to make room for French gray overalls, with a double white stripe, and Wellington boots. These last may perhaps have been introduced rather earlier than the other changes; the Wellington boot, according to one authority, having been prescribed for Light Dragoons in 1808. The old crimson sash of the officer made way for a girdle similar to that worn at present. White welts to the seams and a small pair of epaulettes, white for men and silver for officers, completed the transformation. When the Seventeenth received this new dress it is impossible to say; and the change is therefore recorded under the year when it was ordered, though probably not carried into effect until a year or two later. The fact that the regiment was quartered in India, of course, made in those days no difference as to the clothing issued to it, except that white covers were worn over the shakos. 1812.

In September there arose a mighty famine in Guzerat, which carried off thousands of natives. Simultaneously there broke out an epidemic fever which was as fatal to Europeans as to natives. In the four months, October 1812 to January 1813, four officers and 73 men of the Seventeenth were swept off by this fever; yet even this was a small matter to those who could remember the ravages of yellow fever in the West Indies.

In the three following years strong detachments of the regiment were employed in active service, apparently in expeditions against different hill-tribes. Of the work done I have been unable to discover any record, such expeditions being too common in the early days of British rule in India to excite much interest. In December 1815 the regiment took part in an expedition into the mountains of Cutch, whither no British troops had hitherto penetrated. On the march they crossed the Ran of Cutch, which separates Guzerat from the Cutch peninsula, and being in the advanced guard were the first English soldiers to cross it. The Ran being, from all accounts, merely a bed of sand which comparatively lately had been the bottom of a sea, the accounts of the march and the description of the country filled the Indian newspapers of the period. The news of Waterloo and of the close of the great war was exhausted, so a graphic picture of the Ran was welcome. 1813 to 1815.

The capture of a couple of hill forts, Aujar and Bhooj, soon quieted Cutch; and the troops then repassed the Ran to put down some local banditti and disperse some piratical tribes on the coast. The central nest of these tribes having been taken, the work was done; and accordingly after the capture of Dwarka, on the coast to the south of the Gulf of Cutch, the field force was broken up, and the Seventeenth returned to Ruttapore. 1816.

1816. The losses of the regiment in the work of those three years are unrecorded, and, except from disease, were probably not worth mention.

Before quitting this year we must turn our eyes homeward for a moment, where rather an interesting matter was going forward. H.R.H. the Commander-in-Chief, at the opening of 1816, had become bitten with the notion of forming corps of Lancers in imitation of the Polish Lancers which had done such good service to the army under Napoleon. The first idea was to attach a troop of lancers to each cavalry regiment, just as a small body of riflemen was attached to a regiment of infantry. Lord Rosslyn offered the 9th Light Dragoons for the experiment, and trained fifty picked men under the command of Captain Peters. On Saturday, 20th April, these fifty men were reviewed in the Queen's Riding-house at Pimlico, before a few select spectators who were admitted by ticket. The men were dressed in blue jackets faced with crimson, gray trousers and blue cloth caps, and carried a lance sixteen feet long with a pennon of the Union colours. "The opposite extremity of the lance," continues our authority, "was confined in a leather socket attached to the stirrup, and the lance was supported near the centre by a loose string." Such is an abridged account of the first parade of Lancers in England, taken from an extract from the *Sun* newspaper of 22nd August 1816, and copied into the *Calcutta Gazette*, whence probably it found its way to the officers' mess of the Seventeenth.

1817. The new year brought the regiment to more serious service in the field, namely, the Pindari War. These Pindaris in their early days had been merely the scavengers of the Mahratta armies; but they had been increasing in numbers and power in the south of Hindostan and the north of the Dekhan since 1811. Their most celebrated chiefs were two men named Kurreem and Cheettoo, who had been captured by Dowlat Rao Scindiah, but were released by him for a ransom in 1812. The Pindaris then came out as an independent body, and began incursions on a large scale. They invaded a country in bands of from one to four thousand men apiece, which on reaching the frontier broke up into parties of from two to five hundred. They carried little but their arms; they were admirably mounted, and thought nothing of marching fifty or sixty miles in a day. They lived, themselves and their horses, on plunder, and what they could not carry off they destroyed. In 1812 they were bold and strong enough to cross the Nerbuddha and invade the territory of the Rajah of Nagpore, and in 1813 they actually set fire to part of his capital. As they threatened further depredations in the Gaikwar's territory, a force of 600 native infantry and three troops of the Seventeenth were sent to disperse them; and these repressive measures had a good effect for the time. By 1814 their numbers were reckoned at 27,000 men, "the best cavalry commanded by natives in India," with 24 guns; and in the two following years they became more and more dangerous and troublesome. Holkar and Scindiah, being afraid of them, had both made an alliance with them, and encouraged them secretly. Moreover, the British Government was hampered in any attempt to put them down by an engagement with Scindiah, which prevented it from entering into any negotiations with the Rajpoots under Scindiah's protection. Unless British troops could follow the Pindaris into Rajpoot territory it was of no use to advance against them, for the only way in which the Pindaris could be suppressed was by hunting them down to a man.

1817.

The capture of Bungapore in the Madras Presidency at last brought matters to a crisis. Lord Moira, the Governor-General, called upon Scindiah to disown the Pindaris and conclude a treaty with England. Scindiah signed it cheerfully on the 5th November 1816. That little farce over, he joined a general conspiracy of the Mahratta powers to overthrow British rule in India. The Peishwar and the Rajah of Nagpore, who had also recently signed treaties of alliance with England, together with Holkar were the principal leaders of the movement. Then the Governor-General bestirred himself in earnest. He collected the Bengal,

1817. Madras, and Central armies, and fairly surrounded the whole Pindari country, the Malwa in fact, with 80,000 men. Over and above these a force, under Sir W. Grant Keir, advanced from Bombay to block up one corner on the Bombay side. It was to this force that the Seventeenth was attached, joining it at Baroda.

The Baroda force under Sir W. Keir marched on the 6th December. On the second day's march the rear-guard was attacked by a body of Bheels—a race which, though "diminutive and wretched looking," were "active and capable of great fatigue," as befitted a gang of professed thieves and robbers. They were driven off by a squadron of the Seventeenth under Colonel Stanhope himself, but at the cost of an officer, Cornet Marriott, and several men and horses wounded. Sergeant-Major Hampson received an arrow in the mouth from a Bheel archer. He calmly plucked the arrow out, drew his pistol, shot the Bheel, and then fell dead—choked by the flow of blood. This affair won the Seventeenth the thanks of the General in field orders.

Of the subsequent movements of the Seventeenth in the war I have found great difficulty, from the impossibility of getting at the original despatches, in obtaining any knowledge. The great battle of the campaign was fought against Holkar's troops at Maheidpore on the 20th December. The Seventeenth was not present at the action, though Colonel Stanhope was thanked in orders and despatches for his service as D.Q.M.G., and though immediately after it the regiment was ordered off to reinforce Sir J. Malcolm's division for the pursuit of Holkar. On the 1818. 23rd January 1818 a treaty was make with Holkar; and the war then resolved itself into a pursuit of the other members of the conspiracy, and in particular of the Pindaris. In fact the work of the Seventeenth was a foretaste of that which it was to experience in Central India forty years later; equally difficult to trace from the rapidity of the movements; equally hard to recount from the dearth of material and the separation of the regiment into detachments; above all equally hard on men and horses, perpetually harassed by long forced marches which led only to more forced marches for weeks and weeks together. I have only been able to gather that the men suffered not a little from the extraordinary changes of temperature, varying from $28\frac{1}{2}$ to 110 degrees during the march; and that on a few odd occasions their services were such as to call down the special praise of the divisional commander. These commendations are the more valuable, inasmuch as petty, though brilliant actions were very common in Central India during the early months of 1818.

1818.

The first of these in which we hear of the Seventeenth is an action at Mundapie, wherein four squadrons of the regiment surprised the Pindaris, and cut down 100 of them, with the loss of one private wounded. The gallantry and rapidity of the attack, by the testimony of the General, alone saved the Seventeenth from heavier casualties. We hear next of a detachment of the regiment engaged at the capture of Fort Pallee; and next, at a more important affair, we find the whole of the Seventeenth fighting against the most renowned of the Pindari leaders, Cheettoo himself. The action recalls the history of the detachment which served under Tarleton in Carolina. It appears that Colonel Stanhope obtained information that a large body of Pindaris was within a forced march of him. He at once sent off a detachment in pursuit, which after a thirty mile march came upon the enemy, evidently by surprise, and cut down 200 of them. Cheettoo himself, conspicuous by his dress and black charger, narrowly escaped capture, and owed his safety only to the speed of his horse.[1] Captain Adams and Cornet Marriott, who had already distinguished themselves in the rear-guard action with the Bheels, were prominent on this occasion, and with the whole detachment received Sir W. Keir's thanks in division orders. On the 14th March, when Sir W. Keir's force was broken up, two officers of the Seventeenth, Colonel Stanhope and Captain Thompson, were selected by the General for special approbation and thanks.

19th Jan.

9th Feb.

March.

[1] This animal proved to be Cheettoo's death. His hoofs were so extraordinarily large that his tracks were always recognisable, and hence exposed his rider to the certainty of continued pursuit. Cheettoo having been driven thus into the jungle was finally killed by a tiger.

1819. After a short rest in cantonments the regiment, towards the end of the year, resumed the chase of the Pindaris. The new year found them marching into the province of Candeish, excepting a detachment of eighty-six convalescents who, on their recovery, joined Sir W. Keir's force in Cutch. While there it must have experienced the frightful earthquake of June 1819, which destroyed most of the Cutch towns and killed thousands of natives. Of the general movements of the Seventeenth I have been unable to discover anything. It appears that before the end of the year the regiment was back again in cantonments, and that it moved up to Cutch again in May following, still engaged at the 1820. old work. Colonel Stanhope was then entrusted with a force of between five and six thousand men, destined, it was said, for the invasion of Scinde. After six months' encampment between Bhooj and Mandivie, the Seventeenth returned to cantonments, and the force generally was broken up. Colonel Stanhope, with a few troops which he had retained, reduced the pirate fort of Dwarka, where Cornet Marriott (now promoted Lieutenant in the 67th Foot) was mortally wounded. He was acting as Brigade-Major to Colonel Stanhope at the time, the Seventeenth not being present at the engagement.

Two more years at the Kaira cantonments brought the regiment to the end of its first term of Indian service. It marched to Cambay in November, reached Bombay by water in December, and finally sailed for England on the 9th January 1823. It had landed at Calcutta, in 1808, 790 men strong; it had lost in fourteen years, from disease and climatic causes alone, exclusive of men invalided and killed in action, 26 officers and 796 men; it had received in India 929 men and officers. It went home, after leaving behind it volunteers for different regiments, under 200 strong of all ranks. Such were the effects of cholera,—for 1818 was a bad cholera year,—general ignorance of sanitary matters, and of English clothing in the Indian climate.

GEORGE, LORD BINGHAM
(EARL OF LUCAN)

Lieutenant-Colonel 17th Light Dragoons (Lancers)

1826-1837

CHAPTER XI

HOME SERVICE, 1823-1854

ON their way home the Seventeenth touched at St. Helena, where they found an Army List, and therein learned for the first time that they had become a regiment of Lancers. Such were the fruits of the inspection held at the Queen's Riding-house in Pimlico six years before. There also they heard of the death of their Colonel, Oliver Delancey, who had held that rank since 1795. He had entered the army as a Cornet in the 14th Dragoons in 1766, and joined the Seventeenth as a Captain in 1773. He had therefore held a commission in the regiment for close on fifty years when he died in September 1822. He had gained some slight reputation as a pamphleteer, and he was for many years a Member of Parliament, but it was as a soldier and an officer in the Seventeenth that he had made his mark, in the New England provinces and Carolina. He was succeeded by Lord R. Somerset, a distinguished Peninsula officer. 1823.

On the 18th May the regiment arrived at Gravesend, and marched to Chatham, where all the men, with the exception of some fifty, including non-commissioned officers, were invalided or discharged. At Chatham they returned their carbines into store; it was nearly sixty years before they received them again; and, in accordance with regulation, ceased to shave their upper lips. It must have been rather a curious time, that last half of 1823, between the growing of the moustaches, the learning of the lance exercise, and the constant influx of recruits. In those

1823. days it was, as a rule, rare for a regiment to receive above a dozen recruits in the year; and though the heavy mortality in India had caused the rapid passage of many men into the ranks, yet we may guess that the fifty old soldiers, many of whom had probably brought back with them a liver from the East, were not too well pleased at being flooded with five times their number of recruits. The spectacle of 250 bristly upper lips must in itself have been somewhat disquieting. But recruits came in fast. Before the year was out the regiment numbered 311 men, or little below its reduced establishment, viz. six troops of 335 men with 253 horses.

The acquisition of the lance, of course, brought with it a certain change of dress. Lancers being of Polish origin, the Polish fashion in dress was of course imperative. The shako was discarded for ever, and a lance cap of the orthodox shape introduced in its place; the upper part thereof white as at present, and the plume, as ever since 1759, red and white. The officers, besides a huge pair of epaulettes, wore aiguillettes of silver, and were generally very gorgeously attired. For we are now, it must be remembered, in the reign of King George IV., and therefore every uniform is at its zenith of expense and its nadir of taste. Hence, the first lance caps were so high and heavy that they were a misery to wear; and the jackets, though in pattern unchanged, were made so tight that men could hardly cut the sword exercise.

1824. From this point for the next thirty years the history of the regiment is merely that of home duty in England and Ireland; and as the changes of quarter are recorded in the Appendix, there is no need to repeat them here. Let it, however, be noted that the Seventeenth took the London duty for the first 1825. time in 1824, and that in the following year it found itself once more at Chichester, where we hope that it was welcomed by the Mayor and Corporation.

1826. In 1826, George, Lord Bingham, who had exchanged into the Seventeenth eleven months before, succeeded Colonel Stanhope in command of the regiment. We shall meet with him again as Lord Lucan twenty-eight years hence; not without results. Lord Bingham retained the command until 1837, and brought the regiment up to a very high pitch of efficiency. He was a keen soldier, who had taken the pains to study his profession; a very rare thing in those days; and had even taken the trouble to join the Russian army in the war of 1828-29 against the Turks, in order to gain experience of active service. He came to the Seventeenth at a time when such a commander was especially valuable, for the slack period of the British army, perhaps inevitable after the exertions of the great war, was telling heavily on the cavalry. The drill was stiff, unpractical, and obsolete—designed, apparently, to assimilate the movements of cavalry and infantry as far as possible to each other. It was so useful (this was the pretext alleged) for officers to be able to handle horse and foot with equal facility. "It is hardly credible," writes a critic in 1832, "that the late regulations did not contain a single formation from column into line, in which one or more of the squadrons had not to rein back as a necessary and essential part of the movement." Even when this was altered, officers were still posted in the ranks instead of in front of their troops. At this time, too, and for years after, changes of formation were always carried out to the halt. A regiment that required to take ground to the right, wheeled into "columns of troops to the right," to the halt; then advanced as far as was necessary, then halted, and then wheeled into line, once again to the halt. In many regiments "field cards" were issued, "drawn out in all the pride of red ink," with each movement numbered and marked in its regular succession; and thus the programme for the day of review was rehearsed for weeks beforehand.

1826.

Lord Bingham had not long been in command before the uniform of the regiment was again changed. When the change was made I cannot with accuracy say; but in 1829 we find the white lapel-like facings on the jacket done away with, and a plain blue jacket with white collar and cuffs preferred in its place.

1829.

1829. The old red and white plume also disappears at this period for ever, and a black plume is worn in its stead.

1830. A year later King William IV. came to the throne and made yet another change. Whether from jealousy of the colour of his own service, the Navy, or from whatever cause, he clothed the whole Army, except the artillery and riflemen, in scarlet. The Lancer regiments, one and all, were accordingly arrayed in a double-breasted scarlet jacket with two rows of buttons and gorgeous embroidery, and blue overalls with a double scarlet stripe. The plume for the officers was of black cocktail feathers; and as the cap was very high, and measured ten inches square at the top, and the plume was sixteen inches long, it may be guessed that heads were sufficiently covered. Large gold epaulettes and gold cap-lines with large gold tassels completed the dress. Those were merry days for the army tailor, if not for the Army. That there were curses both loud and deep from the service we need not doubt; but the King at least permitted the Seventeenth to retain its facings, which was more than he allowed to the Navy. With almost incredible want of tact the sailor-king altered the time-honoured white facings of the Navy to scarlet. Happily neither of these changes lasted long; though the appropriation of gold lace to the regular army, and the relegation of silver to the auxiliary forces, has continued to be the rule up to the present day. As a finishing touch to the trials of the Lancers at this period, a general order compelled the shaving of the moustaches which had been so carefully cultivated for the previous eight years.

1828-32. From 1828 to 1832 the Seventeenth was quartered in Ireland. In the latter year they encountered an old Indian enemy in Dublin, namely Asiatic cholera, by which they lost three men. On crossing to England in June they were isolated for some months, lest they should spread the disease from their quarters.

1833. In the following year the regiment was reviewed by King William IV. in Windsor Park. After the review the King invited the officers to dinner, and reminded them then that he

had inspected the Seventeenth half a century before at New York. 1833.
It is noteworthy that one officer, who was still borne on the strength of the regiment, had served with it at that time. Sir Evan Lloyds' name still appeared on the roll as senior lieutenant-colonel ; and thus there was at least one man who could say that he had worn both the scarlet and gold and the scarlet and silver. Nor must we omit to add that among those who witnessed the review on that day was the future colonel-in-chief of the regiment, Prince George of Cambridge, then a boy of fourteen. Thus the lives of two colonels of the Seventeenth actually bridge over the gulf between the American War of Independence and the fifty-eighth year of Queen Victoria. Sir Evan Lloyds' name remained on the regimental list from 1785 until 1836, when he was appointed to the colonelcy of the 7th Dragoon Guards.

1834. The year 1834 witnessed the abolition of a time-honoured institution, namely, the squadron standards. A relic of feudal days, which had kept its significance and its value up to the first years of the great Civil War, the troop or squadron standard had long been obsolete. In fact it is rather surprising that such standards should ever have been issued to Light Dragoons. Nevertheless they survived to a time within the memory of living men in all cavalry regiments, and are fortunately still preserved, together with the blue dress and axes of the farriers and other historic distinctions, in that walking museum of the British cavalry, the Household Brigade.

1837. The year 1837 found the headquarters of the Seventeenth at Coventry for the first time since 1760, when it had but just sprung into existence. On this occasion we may hope that it was allowed to remain in the town during the race meeting. It is somewhat of a coincidence that the regiment should have opened the two longest reigns on record, those, namely, of King George III. and Queen Victoria, in the same quarters. In this same year Lord Bingham retired from the command, and was succeeded by Lieutenant-Colonel Pratt, who in his turn gave place after two years to Lieutenant-Colonel St. Quintin.

1840. In 1840 the Light Dragoons and Lancers discarded the scarlet which had been imposed upon them, and reverted once more to the blue jackets and the overalls of Oxford mixture, which 1841. had been ordained in 1829. In 1841 the Seventeenth, after a three years' stay in Ireland, was moved to Scotland; its first visit to North Britain since 1764. Coming down to Leeds in the 1842. following year it received a new colonel in the person of Prince George of Cambridge, the present Colonel-in-Chief of the regiment and Commander-in-Chief of the Army. Under his command the regiment was employed in aid of the civil power to suppress serious riots in the manufacturing districts in August 1843. 1842. In the following year, headquarters and three troops of the regiment being stationed at Birmingham, there occurred an accident which, after fifty years, sounds almost incredible. The men had just left barracks, in watering order, for the exercise of the horses, and were about to pass under an arch of what in the infancy of railways was called the "Liverpool line," when an engine, with its whistle shrieking loudly, passed over the arch at a high speed. In an instant every horse swung violently round, dismounting almost, if not actually, every man, and the whole hundred of them stampeded wildly back through the streets to their stables. Many of the men were injured, some so seriously that they had to be carried back to barracks; and all this came about through the now familiar whistle of a railway engine. The incident gives us a momentary glimpse of one feature in the England of half a century ago.

1844. Next year the regiment took part in the review held by the Queen in honour of the Czar of Russia. Another ten years was to see it fighting that Czar's army, and helping to break his heart. The vicissitudes of a regiment's life are strange, and the Seventeenth had its share thereof in the forties: first putting down rioters at Leeds; then marching past the Czar at Windsor; then rushing across to Ireland to maintain order there during the 1848. abortive insurrection headed by Smith O'Brien; and, finally, escorting Her Majesty Queen Victoria on her first entry into the city of Dublin. The year 1850 brought it back to England once more, where, after one bout of peace manœuvres at Chobham, it at last received orders, for the first time for thirty-four years, to hold itself in readiness for active service. The warning came in February 1854, and the scene of action was destined to be the Crimea.

1850.

CHAPTER XII

THE CRIMEA, 1854-1856

1854. On receiving the order to prepare for active service the regiment was formed into four service and two depôt troops of the following strength:—

	Field Officers.	Captains.	Subalterns.	Staff.	Sergeants.	Trumpeters.	Farriers.	Corporals.	Privates.	Horses.	
										Officers.	R. & F.
Service	2	4	8	6	18	5	4	13	254	48	249
Depôt	...	1	4	...	7	2	2	5	51	8	34
Total	2	5	12	6	25	7	6	18	305	56	283

April. After the whole had been inspected by the Duke of Cambridge, the depôt troops marched to Brighton on the 10th May, where they formed part of the consolidated cavalry depôt under Colonel Bonham.

Headquarters and the service troops embarked at Portsmouth on the 18th, 23rd, 24th, and 25th April in five sailing ships, thus:—

Headquarters, under Colonel Lawrenson, in the ship *Eveline*.
One troop, under Major Willett, in the *Pride of the Ocean*.
One troop in the *Ganges*.
One troop in the *Blundell*.
Remainder in the *Edmundsbury*.

After passages varying from twenty-three days to five weeks, the whole arrived at Constantinople toward the end of May. Men and officers were all well, but twenty-six horses had perished on the voyage. The regiment was disembarked at Kulali, on the Asiatic side of the Bosporus, and on the 30th of May was inspected by the Sultan in person at Scutari.

1854. May.

On the 2nd June the regiment re-embarked on the same vessel, and sailed to Varna, where, on disembarkation, it was made part of the Light Brigade under the command of Lord Cardigan. Leaving Varna on the 8th it marched to Devna, some eighteen miles to the north-west, and remained encamped at a short distance from the village until the 28th July, on which day it marched for Yeni-bazar, halting at Kutlubi, Yasytepe, and Sazego on the way, and finally encamped at Yeni-bazar on the 1st August. So far the army had done nothing, but had been condemned to inactivity, losing many men by cholera meanwhile. The retreat of the Russians from the Danube after their failure before Silistria, and defeat at Giurgevo in July, had virtually secured the only object of the expedition, namely, that Russia should abandon the invasion of Turkey. But at the end of June the British Government decided to direct the expedition against Sebastopol, and to destroy Russia's great stronghold in the Black Sea. Accordingly, on the 25th of August the Seventeenth started to march back from Yeni-bazar to Varna. Cholera had been at work with them, as with the rest of the army, in August, and they left twelve men buried at Yeni-bazar. Arriving at Varna on the 28th, the regiment embarked once more on four transports on 2nd and 3rd September, and sailed for the Crimea. A fortnight later the headquarters, under Colonel Lawrenson, landed at Kalamita Bay, the spot chosen by Lord Raglan for the disembarkation of the army. The Seventeenth lost two more men by cholera in the passage, and showed a serious falling-off in strength on landing.

4th June.

28th July.

25th Aug.

28th Aug.

17th Sept.

1854.

Field-Officers.	Captains.	Subalterns.	Staff.	Sergeants.	Trumpeters.	Farriers.	Corporals.	Privates.	Total—All ranks.	Horses.	
										Officers.	Troop.
2	4	7	6	16	5	4	11	192	247	21	216

19th Sept. Two days later the army began its advance; the infantry divisions massed in close column, and the cavalry on its skirts—the Seventeenth being in rear of the left flank of the infantry. Early in the afternoon the four squadrons of the advanced guard came upon 2000 of the enemy's cavalry, a little way on the other side of the Bulganak River. Both parties threw out skirmishers, who fired some ineffectual carbine shots without dismounting, as was the fashion of the day; and then the Seventeenth and 8th Hussars were ordered up in haste to reinforce the advanced squadrons. The Russians, although in overwhelming force, did not attack, and the advanced squadrons then retired by alternate wings. A few artillery shots were exchanged, and with that the first encounter with the Russians was over. The troops bivouacked that 20th Sept. night in order of battle, and on the following day attacked and carried the Russian entrenched position on the heights of the Alma.

Details of the action of the Alma, wherein the cavalry, from the nature of the case, was little if at all engaged, would be out of place here. It is, however, worth while to remark that the first infantry division and the cavalry division, which occupied the left of the English line, were both under the command of former colonels of the Seventeenth, the Duke of Cambridge and Lord Lucan. During the infantry attack the cavalry, which was on the extreme left, remained perforce inactive; but when the Highland Brigade, which was next to the cavalry, had carried the heights before them, one squadron of the Seventeenth, which was presently joined by the other, moved off without orders from any general officer, and began to ascend the heights. On their way they contrived in some way to cross part of the front of the Highlanders, and were soundly rated by Sir Colin Campbell for their pains. When, finally, on reaching the summit they began to capture Russian prisoners, the pursuit was checked by Lord Raglan's order; and in consequence little was done. Shortly after the action Colonel Lawrenson went home invalided, leaving to Major Willett the command of the regiment. 1854.

For two days after the battle of the Alma the army remained halted, and then on the 23rd slowly resumed the march on Sebastopol. Lord Raglan's wish had been to push on immediately after the victory, but to this the French commander would not consent. On the 24th the cavalry, under Lord Lucan, was sent on to the river Belbec, a day's march ahead of the main army, but encountered no opposition. Next day, Lord Raglan having been obliged, in deference to the French, to abandon his plan of attacking Sebastopol from the north, the army executed the flank march which brought it round from the north to the south side of the city. The march lay through difficult wooded ground; and the cavalry, which had been pushed forward to cover the advance, was misguided by a staff-officer. The result was that Lord Raglan and his escort were the first to come upon the rear-guard of the Russian army, which was likewise, though unknown to the English, executing a flank march across the British front. The cavalry soon came up, and captured some waggons as well as a few prisoners. After this trifling and rather ludicrous affair with the Russian rear-guard at Mackenzie's Farm, the march was continued, and the army bivouacked that night on the Tchernaya River. On the following day Balaclava was taken; and after three nights more bivouac on the Balaclava plains, the Seventeenth received some tents. They, like the rest of the army, had landed without tents or kits. 23rd Sept 29th Sept.

The main business of the cavalry now consisted in patrolling east and northward towards the Tchernaya, where, as early as the 5th October, it began to encounter Russian patrols. In a sense the cavalry was isolated from the rest of the army. The plain of

1854. Balaclava lies about a mile from Sebastopol, and extends on an average to a length of about three miles from east to west, and a breadth of two miles from north to south. It is enclosed on all sides by heights: on the north by the Fedioukine Hills, on the south by the Kamara Hills, on the east by Mount Hasport, and on the west by the Chersonese, where the bulk of the army was encamped. The plain is cut in two from east to west by a line of hills called the Causeway heights, which run almost to the Chersonese; and it was at the foot of these hills, on the south side of them, that the camp of the Light Brigade was situated. Just about due south of the camp, at a distance of about a mile, stands the village of Kadikoi, at the entrance to the gorge that leads down to Balaclava harbour.

Balaclava was now the British base of operations. Its defence was entrusted to Sir Colin Campbell, with the 93rd Highlanders, some marines, and a certain number of Turks; the cavalry being at hand to help him in the plain. But the better to secure the base with so small a force, an inner line of field-works was constructed from Kadikoi on the north, along the heights on the east of Balaclava to the sea, and an outer line of six redoubts on the Causeway heights. It has already been said that the English and Russian patrols had clashed on the Tchernaya; and as General Liprandi, with a Russian army, had fixed his headquarters at Tchorgoun, less than a mile beyond the Tchernaya to the north-east, this was hardly surprising. Shortly after the middle of October Captain White of the Seventeenth, while on outlying picquet on the Kamara Hills, had observed a large force of Russian cavalry and duly reported it. Knowing the Russians to be in considerable force, neither Sir Colin Campbell nor Lord Lucan were at their ease as to the safety of Balaclava, from the weakness of their defending force and its isolation from the rest of the army.

On the 23rd October Major Willett died, and the command of the regiment once more changed hands. The senior officer, Captain Morris, was employed on the staff; and it became a question whether he would remain where he was, leaving the command to Captain White, or whether he would return to the regiment. On the 24th Lord Lucan received intelligence that Balaclava would be attacked on the morrow by a Russian force of 25,000 men. He at once despatched an aide-de-camp to Lord Raglan, who said "Very well." That evening Captain Morris decided that he would take command of the Seventeenth.

1854.

Next day the cavalry turned out as usual an hour before daybreak, and were standing to their horses, when Lord Lucan rode off slowly to the easternmost redoubt on the Causeway heights. The coming of the dawn showed him a signal on the flagstaff of the redoubt, which told him that his information was correct, and that the Russians were advancing in force. Lord George Paget of the 4th Light Dragoons at once galloped back and ordered the Light Brigade to mount. The men were just about to be dismissed to their breakfasts when they were moved off toward the threatened quarter.

25th Oct.

Meanwhile the Russians, with 11,000 men and 38 guns, attacked the easternmost redoubt; and in spite of a gallant resistance from the five or six hundred Turks that held it, carried it by storm. The Turks then abandoned the three next redoubts; and thus the line of the Causeway heights fell into the hands of the Russians. Simultaneously two more Russian columns had advanced and occupied the Fedioukine heights, and filled the valley between the Fedioukine and Causeway heights with 3500 cavalry and a battery of twelve guns. Lord Lucan, seeing that his 1500 men of the Light and Heavy Cavalry Brigades could not check the advance of 11,000 Russians, fell back to a position on the southern slopes of the Causeway heights, which would enable him to fall on the flank of any force that might cross the South Valley towards Balaclava. From this position he was ordered by Lord Raglan to retire. The result was that the Russians immediately detached four squadrons to attack the weak force of infantry that held the mouth of the gorge leading to Balaclava. So serious did Sir Colin Campbell judge this attack to be that he warned the 93rd, as the Russian cavalry came down on them, that they must

1854. 25th Oct. die where they stood. Fortunately the Russian attack was not pushed home, and the four squadrons were utterly defeated by the unshaken firmness of the 93rd. Convinced as to the soundness of his dispositions, Lord Lucan shortly after moved the Light Brigade forward to its original station; while, in obedience to Raglan's order, he despatched the Heavy Brigade across the valley to reinforce the defending troops at Kadikoi.

Just as the Heavy Brigade was moving off, the Russian cavalry came up in great force over the Causeway heights, full on the flank of the Heavies, but lending their own flank to the Light Brigade. Brigadier Scarlett thereupon wheeled the Heavies into line, and delivered the brilliant attack known as the charge of the Heavy Brigade. Every one, including Lord Lucan, expected to see the Light Brigade fall down on the Russian flank, and smash it completely. But Lord Cardigan judged that his instructions forbade him to attack, and refused to move. Every man in the Brigade was waiting for the order to charge, and Lord Cardigan himself cursed loudly at his own inaction. Captain Morris, doing duty with his regiment for the first time since it had landed in the Crimea, begged and prayed his Brigadier to let loose, if not the whole Brigade, at any rate the Seventeenth Lancers; but Lord Cardigan would not hear of it. Thus for the second time the Seventeenth (and for that matter the Light Brigade), was baulked of the successful attack which its old Colonel had prepared for it.

Then came an order from Lord Raglan to Lord Lucan to "advance and recover the heights," *i.e.* the Causeway heights; presently supplemented by a further order—"Lord Raglan wishes the cavalry to advance rapidly to the front and recover the guns," meaning the guns captured by the Russians in the redoubts on the Causeway heights. This last order was brought by Captain Nolan, an excitable man, and at that particular moment in a highly excited state. "Guns," said Lord Lucan to him, "what guns?" Nolan waved his hand vaguely, it would seem, in the direction of the Russian battery at the head of the North Valley

1854. 25th Oct.

and said, by no means too respectfully : "There, my Lord, is your enemy, there are your guns." Lord Lucan was a quick-tempered man, and probably not in his most amiable mood at that instant. He was one of those officers, rare enough in those days, who had taken particular pains to study his profession, and was on all hands acknowledged to possess more than ordinary ability. His warnings of the previous day had been neglected at headquarters ; his perfectly correct dispositions, carefully concerted with Sir Colin Campbell, had been twice upset by superior order, with results that must almost certainly have been fatal, if the Russian cavalry had known its work ; and now had come a fresh staff-officer with an order which, not in itself too clear, had been further obscured by that staff-officer's excitability. Over hastily he accepted what he believed to be the true meaning of the order, and directed Lord Cardigan to attack the Russian battery at the head of the North Valley with the Light Brigade.

That Brigade, after its various movements, had been finally drawn up facing directly up the South Valley, and had stood dismounted there for more than three-quarters of an hour, when Lord Cardigan gave the order which showed that its time had come. In the Seventeenth that morning there were 139 men in the ranks, increased at the last moment by the arrival of Private Veigh, the regimental butcher, who, hearing that the regiment was about to be engaged, rode up fresh from the shambles to join it. He was dressed in a blood-stained canvas smock, over which he had buckled the belt and accoutrements of one of the Heavy Dragoons who had been killed in the charge ; and, having accommodated himself also with the dead dragoon's horse, he now rode up with his poleaxe[1] at the slope. The rest of the regiment was in marching order — full-dress jackets and lance-caps cased — with the exception of Captain Morris, the commanding officer, who wore a forage cap. The first squadron was led by Captain White, the troop leaders being Captain Hon. Godfrey Morgan and Lieutenant

[1] It is perhaps worth noting that the poleaxe was a favourite weapon with Royalist cavalry officers in the civil war.

1854. 25th Oct.

Thomson; the second squadron was led by Captain Winter, with Captain Webb in command of the right, and Lieutenant Sir William Gordon in command of the left troop. Lieutenant Hartopp, Lieutenant Chadwick (the Adjutant) and Cornet Cleveland were the other officers with the regiment, Cornet Wombwell being with Lord Cardigan as aide-de-camp. The two squadrons of the Seventeenth formed the centre of the first line of the Brigade, having the 11th Hussars to their left, and the 13th Hussars to their right; while the 4th and 8th Hussars composed the second line.

In this formation the Light Brigade moved off to the attack; its duty being to advance over a mile and a half of ground, flanked by Russian batteries and riflemen on the Fedioukine heights to the right, Russian batteries and riflemen on the Causeway heights to the left, and fall upon a battery of twelve guns to their front, which guns were backed by the mass of the Russian Cavalry. The first line began the advance at a trot, and was presently reduced to the Seventeenth and 13th only; the 11th being ordered back to the second line by Lord Lucan. The formation of the Brigade was thus altered from two lines to three. The Seventeenth was now therefore on the left of the first line, though Captain White's squadron still remained the squadron of direction.

Presently, without sound of trumpet, but conforming to the pace of the Brigadier, the first line broke into the gallop. It had barely started when Captain Nolan rode across the front from left to right, shouting and waving his sword. "No, no, Nolan," shouted Captain Morris, "that won't do, we have a long way to go and must be steady." As he spoke a fragment of a shell struck Nolan to the heart. His horse swerved and trotted back through the squadron interval with his rider still firm in the saddle, and then with an unearthly cry the body of Nolan dropped to the ground. This was the first shell that fell into the Light Brigade.

Meanwhile the handful of squadrons, with the Seventeenth and 13th at their head, rode on with perfect steadiness, and in beautiful order, into the ring of the Russian fire. Then men and horses began to drop fast in the first line. The survivors closed up and rode on. The trumpet sounded no charge; the officers uttered no stirring word; the men gave no cheer; for this was no headlong rush of reckless cavaliers, but an orderly advance of disciplined men. Throughout this ride down the valley there was but one word continually repeated, "Close up"; and the men closed in to their centre, and with an ever-diminishing front rode on. Those who watched the advance from the heights a mile away saw the line expand as the stricken men and horses floundered down, and contract once more like some perfect machinery as the survivors took up their dressing and rode on. But at last the gaps became so frequent and so wide that men could close up no more; and then the whole of the first line sat down and raced for the guns. The Russians were ready for them and met them at about eighty yards distance with a simultaneous discharge of every gun in the front battery. How many men fell under this salvo we shall never know. By this time two-thirds of the first line must have fallen: the remaining third rode on. In a few seconds they had plunged into the smoke and were among the Russian guns.

1854.
25th Oct.

On the extreme left a handful of the Seventeenth had outflanked the battery, and of these—all that he could see of his regiment—Captain Morris, who was still unharmed, retained command. Pressing on past the battery through the smoke, he was aware of a large body of Russian cavalry, part of an overwhelming force, that stood halted before him in rear of the guns. Steadying his men for a moment, he led them without thought of hesitation straight at the Russians, and drove his sword to the hilt through the body of their leader. His men were hard at his heels. They broke through the Russian Hussars, they swept all that were covered by their narrow front before them, and galloped on in pursuit. Meanwhile Captain Morris had fallen. Unable to withdraw his sword from the body of the Russian officer, he was tethered by his sword-arm to the corpse, and while

1854. 25th Oct. thus disabled received two sabre cuts and a lance wound. Utterly defenceless against the lances of the Cossacks, who had closed like water upon the small gap made by the Seventeenth, he was forced to surrender. Lieutenant Chadwick, who was wounded by a lance thrust in the neck, was also made prisoner at the same time.

Another fragment of the first line, backed by men of various regiments, was rallied by Corporal Morley, and by him led back through the Russian cavalry to the North Valley.

Yet another little remnant of the Seventeenth, to the right of Morris, had entered the battery, where Sergeant O'Hara took command of them, and directed their efforts against the Russian gunners, who were attempting to carry off their guns. These were presently rallied by Lord Cardigan's Brigade-Major, Major Mayow; but a portion of them having missed him in the smoke went on with O'Hara to their left, where they met their comrades, the survivors of Captain Morris's party. These last, after chasing the Russian Hussars back upon their supports, had been forced back by immensely superior numbers, and were now menaced in their turn both in flank and rear. The two little parties joined together, and fighting their way back through the Russians made good their retreat down the valley.

Meanwhile Major Mayow, with about a dozen men of the Seventeenth, like Captain Morris, charged a body of Russian horse, which was halted in rear of the battery, drove it back, and pursued it for some distance upon the main body. Then Mayow halted, and seeing the remains of a squadron of the 8th Hussars approaching to his right rear, he formed his handful of Lancers on the left flank of the 8th. The Russian cavalry in rear of the guns was now panic-stricken, and in full retreat; but there still remained some Russian squadrons which had been left on the Causeway heights; and of these three now menaced Colonel Shewell's rear. Shewell gave his mixed squadron the word "Right about wheel," and charged them. As usual the Russians received the charge at the halt and were utterly routed. Then, seeing no troops coming to his support, Colonel Shewell retreated.

Once more the British came under the fire of the guns on the Causeway heights. The French had silenced those on the Fedioukine side, the Light Brigade had silenced those in the valley, but those on the Causeway heights still remained untaken. Fortunately some Russian Lancers still hovered about the retreating English, and the Russian gunners ceased to fire lest they should kill their own men. Thus the Seventeenth and the rest of the Brigade returned in small knots well-nigh to the spot from which they had started but five-and-twenty minutes before. Six hundred and seventy-eight of all ranks had started ; one hundred and ninety-five came back.

1854. 25th Oct.

Of the Seventeenth Lancers Captain Winter, Lieutenant Thomson, twenty-two men, and ninety-nine horses were killed. Captain Morris, desperately wounded, finding himself deserted by the Russian officer to whom he had surrendered and left to the tender mercies of the Cossacks, contrived to catch a loose horse, and, when this had been killed under him, made shift to stagger back to the place where Captain Nolan had fallen. There he dropped, but was tended under fire by Surgeon Mouat and by Sergeant Wooden of the Seventeenth, both of whom received the Victoria Cross for the service. Captain Robert White was badly wounded before reaching the battery, and Captain Webb wounded to the death. Sir William Gordon, who had passed through the battery unharmed, came back from pursuing the Russian cavalry with five sabre wounds in the head. So terribly had he been hacked that the doctors said that on the 25th October he was "their only patient with his head off." Hardly able to keep himself in the saddle he lay on his horse's neck, trying to keep the blood out of his eyes, and rode back down the valley at a walk. Being intercepted by a body of Russian cavalry he made for the squadron interval, followed by two or three men, and when the Russians, in their endeavour to bar his passage, left an opening in the squadron, he managed to canter through it and in spite of pursuit to finally complete his escape. His horse, which was shot through the shoulders, managed to carry

1854. 25th Oct.

him out of action, but died, poor gallant beast, very soon after. Thirty-three men and almost every surviving horse were also wounded; Trumpeter Brittain, who had acted as Lord Cardigan's trumpeter on that day, dying of his hurts in hospital. Lieutenant Chadwick, and thirteen more men, all of them wounded, were taken prisoners. Lieutenant Wombwell, being like Captain Morris abandoned by his captors to the Cossacks, escaped, after having two horses killed under him.

So ended the work of the Seventeenth on the 25th October 1854. It is customary to look upon the attack of the Light Brigade as a mere desperate ride into the Russian battery. It was far more than this. The advance down the valley through the murderous fire from front and both flanks was but the prelude to a brilliant attack. Discipline never failed even among the scattered fragments of the first line. Where their own officers were still alive with them, the men of the Seventeenth, however trifling in numbers, rallied, as under Captain Morris, and followed them to the attack on the Russian cavalry. Where an officer of another corps rallied them, they followed him with the same devotion and intrepidity. The little knot with Major Mayow, under his leadership attacked ten or fifteen times their number of Russians, defeated them, pursued them, halted, rallied on the 8th Hussars, attacked with them successfully once more, and stood ready to renew the attack yet again if supports should come. Where, again, no officer was present, the non-commissioned officers, true to regimental tradition, readily took command; and Sergeant O'Hara and Corporal Morley proved themselves worthy successors of Tucker and Stephenson.

Had the attack of the Light Brigade been supported there is reason to suppose that it would have been brilliantly successful; for the Russian cavalry had been thoroughly scared, and even the infantry had been formed into squares to resist the onslaught of the few score of men who had passed the battery. Lord Lucan had indeed every intention of supporting it with the Heavy Brigade, and actually brought that brigade within destructive fire;

but seeing from his advanced position up the valley the frightful losses of the Light Brigade, he could not bring himself to sacrifice the Heavies also. Pulling up under the cross-fire of the batteries, his horse wounded in two places, and his own thigh injured by a musket ball, he took his resolution and ordered the Heavy Brigade to retire. What his feelings may have been when he saw the wreck of his old regiment return to him we can only guess. Yet this was not the first occasion on which the Seventeenth had charged ten times their number of cavalry; they had done it once before at Cowpens against a far more dangerous and resolute enemy.

1854.

After Balaclava the Seventeenth, like the other four regiments of the Light Brigade, had almost ceased to exist in the Crimea, from the extent of its loss both in men and horses. A supply of remounts was, however, obtained by the capture of about 100 Russian troop-horses which stampeded into the British camp on the night of the 26th October.

5th Nov.

The next great action of the war was the battle of Inkermann on the 5th November. In this engagement the brunt of the work fell, from the nature of the case, upon the infantry. The Light Brigade was, however, brought under fire late in the day in support of some French reinforcements; Lord George Paget, who was in command that day, having received instructions, and also a particularly urgent request from the Commander-in-Chief of the French, to keep his men, a bare 200 all told, within supporting distance of the French cavalry. The losses of the Light Brigade amounted to an officer and five men killed, and five men wounded, of whom the officer and another of the killed and one of the wounded belonged to the Seventeenth. Cornet Cleveland, who had escaped at Balaclava where so many fell, was the only English cavalry officer who was touched at Inkermann. His death reduced the number of unwounded officers of the regiment to three.

25th Nov.

Three weeks later the establishment of the Seventeenth was raised to eight troops—a curious reflection for the handful

1854. of men who represented it in the Crimea. Some months were yet to pass before the Seventeenth at Sebastopol could make any show as a regiment, and those months were those of the Crimean winter. So much has been written of that terrible time that it would be out of place to say much of it here. Suffice it that between bad luck and bad management both men and horses suffered very severely. Probably there never was a time excepting the winter of 1854 when the troop-horses of a British cavalry division were almost without exception hog-maned and rat-tailed, the poor creatures having eaten each other's hair in the extremity of hunger. As to the men of the Seventeenth, it is enough to say that they shared the misery and hardship which was borne by the rest of the army, which was cruel enough. But hard as was the Crimean winter, it must not be treated, simply because a British war-correspondent was present and a British Parliament was busy, as an unique trial of endurance. A regiment which had fought through the Carolina campaigns and the deadly war in the West Indies had little new to learn of misery, sickness, and death.

1855. In the months of April and June of the following year the regiment received large drafts from England, and by the 21st July was enabled to detach a squadron of 100 men and horses, under the command of Captain Learmonth, to join a force of British cavalry which was employed in collecting forage and supporting the French in the Baidar Valley. This squadron rejoined headquarters on the 19th August, in time to be present together with the rest of the regiment at the battle of the Tchernaya. Three weeks later Sebastopol was evacuated, and the war was practically over.

20th Aug.

8th Sept.

About the middle of November the regiment embarked at Balaclava for Ismid, where it landed on the 15th. Its strength on embarkation was 15 officers and 291 non-commissioned officers and men, with 224 horses; and the whole of it was carried in two transports, the *Candia* and *Etna*. A corporal and five men were left behind to do orderly work in the Crimea. At Ismid the Seventeenth was brigaded with the 8th and 10th Hussars, under Brigadier Shewell, and there remained until after the proclamation of peace. 1856. 30th Mar.

On the 27th of April a sergeant's party of seventeen men and sixteen horses was embarked in the transport *Oneida*, and two days later the bulk of the regiment, 18 officers and 442 men, with 171 horses, embarked in the *Candia*, homeward bound. The whole arrived at Queenstown on the 14th May, having suffered no casualty but the loss of a single horse on the passage.

On landing, the regiment was quartered at Cahir barracks (where it was joined by the depôt squadron from Brighton), with detachments at Clogheen, Clonmel, Fethard, and Limerick. It had not been at home two months before it was employed at Nenagh in aid of the civil power. In September the regiment was moved up to Portobello Barracks in Dublin, and two months later was reduced to six troops once more, with an establishment of 28 officers, 442 non-commissioned officers and men, with 300 troop-horses. Early in the following year it moved to Island Bridge Barracks, where all the elaborate arrangements for quarters and reduction of establishment were upset by the outbreak of the Indian Mutiny. 12th Sept. 10th Nov. 1857. 7th Mar.

CHAPTER XIII

CENTRAL INDIA, 1858-1859

1857. FOR the better understanding of the share taken by the Seventeenth Lancers in the suppression of the Indian Mutiny, it may be well to set down as briefly as possible the principal events that had taken place before their arrival—

First outbreak at Meerut . . .	10th May 1857.
Outbreak at Lucknow	30th " "
" " Cawnpore . . .	7th June "
Siege of Delhi opened . . .	8th " "
Cawnpore massacre	26th " "
Capture of Cawnpore by Havelock . .	18th July "
Fall of Delhi	20th Sept. "
First relief of Lucknow . . .	25th " "
Second " " . . .	17th Nov. "

In those days, when there was neither submarine cable nor Suez Canal, news from India took some time to reach England. Reinforcements destined for China were intercepted and sent to India on their way, and thus arrived early; but it was October 1857 before the reinforcements from England began fairly to pour into Calcutta. The Seventeenth was not of these first reinforcements; and did not receive its orders for embarkation before 2nd September. On the 7th of that month its establishment was raised from six to ten troops; and volunteers, to the number of 132, were received from other regiments, namely the 3rd, 4th, and 13th Light Dragoons, the 11th Hussars, and the 16th Lancers. It will be noticed at once that this list includes three regiments out of the five which had composed the Light Brigade in the Crimea. The other regiment of that Brigade, the 8th Hussars, sailed with the Seventeenth to India. 1857.

On the 1st October the depôt was formed, and on the 6th the regiment moved by rail from Dublin to Cork and embarked on board the steamship *Great Britain*, wherein the 8th Hussars had already been embarked on the previous day. The strength of the Seventeenth was as follows:—

Field Officers.	Captains.	Subalterns.	Staff.	Sergeants.	Trumpeters.	Farriers.	Corporals.	Privates.
3	4	9	5	37	6	8	23	409

We may note among the officers the names of Captains White and Sir W. Gordon, whom we knew at Balaclava, and of Captain Drury Lowe and Lieutenant Evelyn Wood, whom we are in future to know better.

On the 8th October the *Great Britain* sailed, and after touching at the Cape de Verdes and the Cape of Good Hope to coal, reached Bombay on the 17th December. A single casualty, the death of a private from heart disease, alone occurred on the seventy days' voyage. The Colonel, who with one captain, the riding-master, the veterinary surgeon, and four rough-riders, had been sent out by the overland route, of course reached India earlier than the rest of the regiment. The Seventeenth disembarked in two divisions on the 19th and 21st December, and on landing were moved up first to Campoolee, at the foot of the Bhore Ghauts, and thence to Kirkee cantonments, where it arrived on the 24th and 26th.

Then came a weary period of waiting until horses could be procured from the remount establishment in Bombay. Meanwhile, on the 6th January 1858, Sir Hugh Rose opened the extraordinary campaign wherein he marched from Indore, and fought his way without a check to the Jumna. But when he had closed this campaign, first at Calpee on the 24th May, and finally at 1858.

1858. Gwalior on the 20th June, the most strenuous of his enemies were still at large, and, as the event proved, not to be captured for another nine months. These were Tantia Topee and the Rao Sahib; the latter Nana Sahib's nephew, the former his right-hand man. Of the two Tantia was incomparably the more formidable. After being present at the first siege of Cawnpore, and the subsequent defeat of the Nana's troops by Havelock, he had been entrusted with the command of the Nana's "Gwalior contingent." With this he had beaten General Wyndham before Cawnpore (26th and 27th November 1858), and though immediately after defeated in his turn by Sir Colin Campbell, had by no means abandoned the struggle. Turning north from Cawnpore he first captured Chirkaree. He then tried to relieve Jhansi, at that time besieged by Sir Hugh Rose, and was defeated (1st April 1857); and meeting Sir Hugh Rose once more at Kunch, was again defeated. Still unquelled, he turned against Gwalior, routed Scindia's troops, and captured the fortress. There he was for the third time defeated by Sir Hugh Rose, and his force still further dispersed by Sir R. Napier at Jowra Alipore (22nd June). He then tried to make his way northward, but was headed back by General Showers. Still undismayed, he broke away westward to Tonk; from which point begins the final act of the drama of the Mutiny. In this act, which may be called the hunting of Tantia Topee, the Seventeenth had its part, and played it on the old stage of the Pindari war—Malwa.

While Sir Hugh Rose was fighting, horses began to arrive at Kirkee—Arab, Syrian, Australian, and Cape horses for the most part; and as each squadron of the Seventeenth was mounted, it was hurried up to the front to join in the chase of Tantia. The first squadron was despatched from Kirkee on the 27th May, under the command of Captain Sir William Gordon, to join Major-General Michel's force at Mhow. This squadron, in spite of many obstacles, lost no time upon the road. The first difficulty was the desertion, after two or three days' march, of the *baboo* who was in charge of the Commissariat arrangements. His place was taken by the only officer who could speak Hindustani, Lieutenant Evelyn Wood; and the squadron marched on without a day's halt for the whole of the five hundred miles to its destination, learning much on the way, and arriving in perfect condition. At whatever hour of the day or night the march might close, Sir William Gordon, with or without the help of a candle, inspected every horse's back, and if the hair appeared to be in the least degree ruffled, shifted the stuffing of the saddle from the tender place with a homely but effective instrument, a two-pronged steel fork. If the back were actually sore the trooper could look forward to the pleasure of tramping with the rear-guard on his own feet until it was healed; for this was the "golden rule" from which the Captain never departed. And such a tramp was not altogether enjoyable at that season. On the day before the squadron ascended the table-land whereon Mhow stands, the heat was so intense that the backs came off the brushes, and the combs contorted themselves into serpentine shapes. But there was not a sore back in the squadron when, at the end of June, it reached its destination, nor through the whole of the arduous service that subsequently fell upon it.

1858.

By that time Tantia had already travelled over a large extent of country. Closely followed by two flying columns under General Roberts and Colonel Holmes, he struck southward from Tonk, and was overtaken and defeated by Roberts at Sanganir on the 7th August. A week later (14th August) he was again attacked by Roberts at Kankrowlee, again defeated, and pursued for seventeen miles. Then he struck east towards the Chumbul, where he evaded a third column under Brigadier Parke and reached Jhalra-patan. Here he was joined by the Rajah's troops, whereby his force was augmented to 10,000 men, and gained possession of forty cannon as well as of considerable treasure.

Thus strengthened, he conceived the idea of marching on Indore; but General Michel, divining his purpose, sent two columns, under Colonels Hope and Lockhart, to cut him off. Tantia then retired leisurely to Rajghur. General Michel there-

1858. upon moved up to Nulkeera, about a hundred miles north of Mhow, and there added his troops, including Sir W. Gordon's squadron of the Seventeenth, to the united columns of Colonels September. Hope and Lockhart. On the 14th September Michel, having obtained information of Tantia's movements, marched on Rajghur, some five-and-thirty miles distant.

His force consisted of the following troops :—

Seventeenth Lancers	80
3rd Light Cavalry	180
71st Highland Light Infantry and 92nd Highlanders	600
15th and 4th Rifles, N. I. / 4 guns, Bengal Artillery	240
	1100

Heavy rain was falling, and the cotton soil of Malwa was a sea of black mud. With great difficulty Michel reached Chapera, about half-way to Rajghur, and there halted. Next day the rain ceased, and the heat was so terrible that one-third of the European infantry fell out exhausted, several of them actually dying of sunstroke, while many of the artillery horses dropped dead in the traces. The march that day lasted from 4 A.M. till 5 P.M., when Michel at last arrived in sight of the enemy; but his infantry were then three miles in rear of the mounted men, and so much spent that attack was out of the question.

At 2.30 next morning Michel advanced, but found that Tantia had retired. The Seventeenth and the native cavalry, the whole being under the command of Sir W. Gordon, were pushed forward on the track of Tantia's retreat, and presently came upon his whole force, 8000 men and 27 guns, drawn up for battle in two lines. After a trifling skirmish the cavalry was halted to permit the infantry and guns to come up; but the rebel army, on seeing the advance of the British, forthwith gave way and fled. Then Sir W. Gordon, who had been posted on the extreme right, was let loose with the cavalry, and dashing to the front, dispersed (to use Michel's own words) all symptoms of an organised body. The pursuit was kept up for four or five miles till men and horses were tired out. The heat was terrible; the infantry fell out in great numbers under the midday sun; and when the cavalry finally halted under the shade of some trees, an officer of the native cavalry died then and there from sunstroke. But not a drop of blood was shed on the English side; and the losses of the Seventeenth consisted of a single horse killed. The trophies of the cavalry consisted of Tantia's whole park of 27 guns.

1858. 15th Sept.

After one day's halt Michel resumed the pursuit, passing eastward through Nursinghur; but between that place and Birseeah the rain came down with such violence that further progress was impossible. For two days the torrent never ceased to fall. The camp became a swamp, and the unfortunate horses stood fetlock deep in mud. Meanwhile Tantia moved away through dense jungle to the north-eastward, and on reaching Seronge, fifty miles from Rajghur, halted there for eight days. He then moved northward sixty miles to Esaughur, one of Scindia's forts, which he stormed and plundered, capturing some supplies and seven guns. He used one of these guns for the purpose of blowing his chief artillery officer from its mouth, and then took counsel with the Rao Sahib as to future operations. The pair then agreed to divide their forces—Tantia moving eastward to Chunderi, and the Rao Sahib northward to Tal Bahat.

After wasting three days in the vain attempt to capture Chunderi from Scindia's garrison, Tantia moved south about twenty miles to Mungrowlee—as fate ordained it, straight into the jaws of his pursuers. Michel having marched since daybreak thirty-five miles north-eastward from Seronge, was in the act of pitching his camp at Mungrowlee, when a lancer of the picquet galloped in with the report that the rebels were close at hand. Michel's force was made up as follows:—

Seventeenth Lancers	90
H.M. 71st and 92nd	510
19th N. I.	429
Bengal Artillery, 4 guns . . .	62
	1091

1858. 9th Oct.

Tantia Topee had 5000 men and 6 guns. His advanced guard alone was visible when Michel moved out to meet him, and he himself was quite unaware of Michel's proximity. Tantia's position, as it happened, was strong; his advanced guard having reached an elevated village, surrounded by high scrubby jungle, in which it was impossible for infantry to perceive an enemy, while his guns commanded the ground over which the British must advance. With unusual boldness Tantia sent his cavalry forward and menaced both flanks of the British. Just at that moment an alarm was raised in the British rear. A party of Velliattees had contrived, owing to the thickness of the jungle, to steal up unperceived in rear of Michel's support, and had succeeded in murdering a wounded Highlander. Sir W. Gordon at once galloped up with his troop of the Seventeenth; whereupon the Velliattees promptly vanished into the jungle. With some difficulty Sir W. Gordon espied some of their heads through the foliage, and forthwith gave the order to open out and pursue at the gallop. In an instant the handful of men dashed into the jungle, heedless of what might be there, and was in the midst of the Velliattees. Order of any kind on such ground was impossible, so every man worked for himself; and with such effect did the lances play that when the Seventeenth finally emerged from the jungle they left over eighty of the rebels dead on the ground. Every man of the forty-three that were present of Sir William Gordon's troop killed two, and Gordon himself, galloping like the wind, killed four with his own sword, and knocked over as many more with his horse's chest. He had, however, a narrow escape; a rebel, who was just about to fire at his back, being killed in the nick of time by Sergeant Cope. Tantia's main army as usual turned and fled when the British infantry fairly advanced against them. Had Michel's cavalry been more numerous he might have cut the whole of the rebels to pieces; but, as things were, he had to be content with one hundred of them left dead on the field, a large number of prisoners, and Tantia's six guns. "I solicit to bring Sir William Gordon's services prominently to the notice of His Excellency," wrote General Michel after this action, "and those of the squadron under his command, who did their duty admirably."

1858.

After his defeat at Mungrowlee Tàntia fled eastward across the Betwah to Lullutpore, where he rejoined the Raö Sahib. There he remained while the Rao Sahib marched eastward with 10,000 men and six guns. General Michel meanwhile divided his force into three columns, intending to move himself with the centre column in a direction due east; but finding that his intended route lay through jungle infested by predatory tribes, he made forced marches southward in order to join with his right or southern column once more. Overtaking this column at Narut on the 18th October he had ordered a march north-westward towards Lullutpore, when at 1 A.M. he received intelligence of the presence of the Rao Sahib at Sindwaho, fifteen miles to the north. In an hour Michel had started to meet the enemy, and at daybreak his cavalry came into sight of one of the rebel picquets close to Sindwaho. His force was composed thus:—

18th Oct.

19th Oct.

R. H. A. (4 guns)	68	71st Highland Light Infantry	210
8th Hussars	118	92nd Highlanders	320
Seventeenth Lancers	90	19th N. I.	500
1st Bombay Lancers	93	Bengal Artillery (4 guns)	60
3rd Bombay Cavalry	98	3rd Bombay Cavalry	50
Mayne's Horse	150		
	617		1140

The village of Sindwaho lies between the Jamnee river and its tributary the Sujnam. The country round it has a general elevation of about fifteen hundred feet, with an undulating surface broken by numerous detached hills and peaks. There is very little cultivation on the high land, the greater part thereof being covered with dense jungle. The Rao Sahib had drawn up his force, 10,000 strong, on rising ground, and so disposed it as to conceal his exact numbers. His artillery was just over the sky-line, with cavalry on either flank, and some squares of infantry in the jungle, which here and there was partly open. He awaited

1858. attack, having sent down to the edge of a watercourse detached bodies of infantry to annoy Michel's force as it went into the broken ground at the bottom.

Michel at once sent off the cavalry to his extreme right in order to cut off the enemy from their ascertained destination. By chance the rebel artillery found the range of the British at once, and by three or four lucky shots caused some slight loss to the Seventeenth while executing this movement. The English guns, with a strong escort, occupied Michel's centre. As at Mungrowlee, the rebels made a show of taking the initiative, their infantry advancing against the guns while their horse hovered about the flank of the British cavalry, which charged them with great effect. Then Michel's infantry came up, and was actually so far pressed by the enemy that one flank needed to be reinforced, while the artillery in the centre was obliged to fire grape. But as usual the rebels did not stand long; and presently Sir William Gordon, with the Seventeenth, the 8th, and the Bombay Lancers was in the thick of them. For nine miles the pursuit was continued, though, from the heavy condition of the cultivated land and the broken nature of the ground, it was inevitably slow. None the less 500 dead rebels and 6 captured guns made the victory tolerably complete.

While the bulk of the cavalry was thus engaged on the right, an escort of the 3rd Bombay Cavalry, in attendance on a couple of guns on the left, was fired at by a small body of rebels from a field of high *jowarree*. Several horses having been wounded, the escort was withdrawn for a little distance; and thereupon these rebels, many of whom were mutinous Sepoys of the 36th Bengal Native Infantry, drew themselves up into a kind of rude square. Lieutenant Evelyn Wood of the Seventeenth, who had been doing duty with the 3rd Light Cavalry since they left Mhow, no sooner saw this square than he attacked it singly and alone, selecting the corner man as his first opponent. While he was engaged with him a sowar of the 3rd Light Cavalry, Dokal[1] Singh, came up,

[1] Now A.D.C. to the Governor of Bombay.

and, having narrowly escaped a cut from a two-handed sword which shore through his saddle into his horse's spine, presently made an end of the corner man. Then a small party of the 8th Hussars, under the Adjutant, Mr. Harding, was brought up to Lieutenant Wood's assistance by Lieutenant Bainbridge of the Seventeenth, and the rebels began to disperse. Harding called out to Wood to fight one of them, and himself selected another. The sepoy waited for Harding until he was so close that the fire of the musket singed his stable jacket, and shot him dead. Lieutenant Wood's opponent also waited for him with the bayonet, till finding the chest of his horse almost on the top of him, he clubbed his musket and was at once run through the body by Wood's sword. This was one of two gallant actions for which Lieutenant Wood (better known as Sir Evelyn Wood) received the Victoria Cross. 1858.

For the rest the rebels made a better resistance in this action of Sindwaho than in any other of the many that were fought during the chase of Tantia. The total loss of the British did not exceed 5 officers and 20 men killed and wounded; but the brunt of the day's work and the whole of the loss fell on the cavalry. Of the Seventeenth one sergeant and four privates were wounded; three horses killed and four wounded. Sir William Gordon was again honourably mentioned in despatches; and Lieutenant Wood distinguished himself as has been already told. The cavalry, when the day's work was done, had been in the saddle from 2 A.M. till 5 P.M., and was not sorry to rest. Still, they had more than ordinary consolation, for on one native saddle were found gold mohurs to the value of £150, which were distributed among the men. Let us not omit to mention, also, that the infantry almost kept up with them during the twenty mile march that preceded the action, and that among the infantry regiments, in this as in the two previous engagements, was the 71st Highland Light Infantry, which had worked through so many hard marches with the Seventeenth in the Carolinas three-quarters of a century before.

After one day's halt General Michel marched from Sindwaho

1858. northward to Lullutpore. Then Tantia made a desperate move. Starting from the northward of Lullutpore he doubled back suddenly to the south, passing unobserved within four miles of the British column, and between it and the Betwah. Michel, on learning of this new departure, instantly followed him by forced marches from Lullutpore; but being unable to pursue him directly by the mountains and jungly track that Tantia had selected, he was compelled to move by Malthor (a thirty mile march) and Khimlassa, where on the evening of the 24th he heard that Tantia had but just passed before him. On the 25th at 2 A.M. Michel resumed the pursuit, and at Kurai overtook the wing of Tantia's army, 2000 strong. This force made hardly even a show of fighting, but forthwith fled and was hotly pursued by the British cavalry in three separate columns. Sir W. Gordon, with the Seventeenth and the 3rd Light Cavalry, pressed the rebels hard for six miles, and as usual (to quote General Michel's despatch) did his work efficiently and well. In the course of the pursuit, while hastening with all speed after some cavalry that was covering the retreat of some rebel leader, the Seventeenth were brought up, as is so often the case in that country, by a nullah. Sir William Gordon, as was, of course, his invariable rule, waited until he had seen every trooper pass over before him, and then gave the word to open out and pursue at the gallop, adding that the first man up should have for his reward whatever the leader carried on him. Well mounted, and an admirable horseman, Sir William won the race, killed the leader with his own hand, and divided the gold bracelets and other ornaments of great value that were on his body among the men that were first after him. It is hardly surprising that his troop did wonders under such a Captain. Let us, however, do justice to all, and record the extraordinary marches accomplished by the infantry of the column just at this time—twenty-nine miles on one day, twenty-seven on the next, and twenty-five before they came into action at Kurai.

25th Oct.

The wing thus caught by Michel was simply dispersed; and

(in the words of the historian of the Mutiny) Tantia and the Rao Sahib purchased their retreat by the sacrifice of one-half of their followers. 1858.

None the less Tantia pushed on with such force as he had saved. He was again attacked on the following day by a single regiment—that now known as the Central India Horse—and suffered some loss; but still he pushed on. Within a few days he had crossed the Nerbuddha, to the great alarm of the Governments at Madras and Bombay, and was pointing towards Nagpore.

Headed back from thence by a British force, he turned sharp to the west, hoping to find some unguarded pass by which he might pierce farther south. It was useless; every outlet to south and west was already occupied. He then turned north-westward into Holkar's country, forced a certain number of Holkar's troops to join him at Kargun (19th November), and then hurried away towards the west. November.

Meanwhile Michel had followed him across the Nerbuddha, reaching Hoshangabad on the 7th November. Feeling sure of the security of the south and west, he sent Brigadier Parke on to Charwah, and followed in the same direction more leisurely himself. Sir William Gordon's squadron was left for a time at Hoshangabad, where it was presently joined by further portions of the Seventeenth. It is now necessary to pause for a moment and go back to the rest of the regiment, which we left at Kirkee awaiting its establishment of horses.

The second squadron, under Major White, left Kirkee on the 11th June and marched to Sholapore, where it was kept halted for some time. We shall, however, see this squadron in action in due season.

The third squadron, under Major Learmonth, left Kirkee on the 11th September, and proceeded to Mhow, where it was placed at the disposal of General Michel.

Headquarters and the remaining squadron, having left a small depôt at Kirkee, marched from that station on 22nd September, in company with D troop of the Royal Horse Artillery and some

1858. November.

infantry, the whole being under the command of Colonel Benson of the Seventeenth. On arrival at Mhow they were immediately pushed forward towards the Betwah, and having picked up first Major Learmonth's squadron at Bhopal, and next Sir William Gordon's at Hoshangabad, united three-fourths of the regiment at the latter place on the 6th November.

Meanwhile Tantia was still pressing on with all speed to westward. On the 23rd November he crossed the great high-road from Bombay to Agra, plundered some carts laden with mercantile stores for the army, cut the telegraph wires, and hurried on in the hope of recrossing the Nerbuddha unperceived. The British were quickly on his track. Major Sutherland, with a handful of 200 infantry, caught him at Rajpore, attacked him, though against odds of fifteen or twenty men to one, and put him to flight. Nevertheless, though the pursuit was resumed next morning with all possible swiftness, it was only to find that Tantia was safe across the Nerbuddha. Tantia then moved rapidly north in the hope of surprising Baroda; but the British were beforehand with him. Brigadier Parke, moving by extraordinary marches, met him at Oodeypore on the 30th of November and defeated him once more. Tantia then fled eastward into the Banswarra jungle, and the British commanders thought that they had caught him at last. He was not caught yet by any means. The next that the Seventeenth heard of him was that he was advancing on Indore, and that they must move up to Mhow with all speed. Colonel Benson left his encampment, twelve miles south of the Nerbuddha, crossed the river in boats, and was at Mhow in twenty-six hours—a march of fifty-two miles, to say nothing of the passage of the river.

Tantia, however, prudently remained in the jungle; and on the 3rd December Colonel Benson, with his three squadrons of the Seventeenth, again left Mhow and marched north-westward for Ratlam, in order to meet him whenever he might issue from his hiding-place. A small column under Major Learmonth was detached from Ratlam, but after three days' search discovered nothing

December.

of the enemy ; and Colonels Benson and Somerset, who had united 1858. their two flying columns at Ratlam, then moved up together to Partabghur. At this point, however, a new ally for Tantia, Feroz Shah, appeared upon the scene, and Somerset's column was detached to Ashta to cut him off. Emboldened by Feroz Shah's diversion, Tantia finally emerged from the jungle, after a month's wandering, at Partabghur, on Christmas day 1858. But meanwhile Colonel Benson had been moved from Partabghur; and a very weak force of native infantry alone was on the spot to stop the famous rebel. Tantia held this little force engaged for a couple of hours until his baggage and elephants were clear of the passes, and then marched quietly away. Halting for the night within six miles of Mundesoor he struck eastward, and in three days had reached Zeerapore, one hundred and ten miles as the crow flies from Partabghur.

Meanwhile Colonel Benson had lost no time in starting on his track with 210 men of the Seventeenth and 37 men of the Horse Artillery with 2 guns; and after a march of one hundred and forty-eight miles in one hundred and twenty hours, he finally caught Tantia at Zeerapore. This being, so to speak, a strictly regimental affair, we may give an abridged journal of the march :—

Friday, 24th December.—Left Ninose for Nowgaum (seventeen miles).

Saturday, 25th December.—Made a reconnaissance, and discovered that the enemy had marched on Mundesoor ; made a forced march thither, and arrived that night (thirty-six miles) to find the enemy encamped but four miles away.

Sunday, 26th December.—Marched at daybreak, leaving behind all infantry, artillery waggons, led horses, and baggage of every description, and all grass-cutters. Moved first towards Seeta Mhow on false information, but, discovering the true direction, turned towards Caimpore, and halted for the night on the left bank of the Chumbul (twenty-six miles).

Monday, 27th December.—Marched at daybreak, crossed the Chumbul, and came up with the rebels encamped at Dug ; bivouacked in sight of their fires.

Tuesday, 28th December.—Marched at 4 A.M. so as to attack at day-

1858. break; found that the enemy's main body had retreated. Drove in the picquets and pursued, crossing the Kollee Sind River on the way (twenty-eight miles).

Wednesday, 29*th December*.—Marched at 3 A.M. from the right bank of the Kollee Sind; after an eight-mile march came in sight of the rebel camp; advanced over the ploughed land, so as to make as little noise as possible, and waited for daylight. Found the main body had retired two miles; trotted on and came up with it; and on emerging from a wooded lane found the rebel army, apparently about 4000 strong, drawn up in line of battle on rising ground, with a ravine and jungle to their rear.

29th Dec. Colonel Benson advanced to the attack in columns of divisions, and, on the commencement of the rebel fire, moved the leading column to the right, thus uncovering his guns, which opened fire at four hundred yards with grape and shell. The rebels soon gave way, and Benson then attacked with two divisions from his right, and drove them into the jungle. The Seventeenth then pursued them through the jungle and across the ravine, and on emerging from the latter found them rallied and drawn up in a new position. The Seventeenth then advanced in line, with the two guns in the centre, and after a vain attempt of the rebels to make a counter-attack, Sir William Gordon charged with his squadron and drove the enemy once more into the jungle and across the ravine. With some difficulty and delay the guns were taken across in pursuit; and after one or two more feeble attempts to rally, the rebels were dispersed and pursued in all directions. The action closed with the capture of four of Tantia's elephants by Captain Drury Lowe. The ornaments of these elephants still remain in the regiment's possession as trophies of this regimental day. The whole affair lasted about two hours; and the distance covered before the day's work was ended was thirty-six miles, making a total of one hundred and seventy-eight miles, including the passage of two large rivers, in six days, accomplished without European supplies, without protection against the bitter cold of the nights, and, above all, without a murmur. The casualties were as usual trifling enough. The Artillery and Seventeenth each lost one man wounded and two horses killed. 1858.

On the very next day (30th December) Colonel Somerset's column, consisting of 4 guns of the Royal Horse Artillery, 100 of the Seventeenth under Major White, and 150 of the 92nd Highlanders on camels, arrived likewise at Zeerapore. Major White had just missed Colonel Benson at Dug by three hours; and had then been summoned to join Colonel Somerset at Soosneer. In consequence of information as to a junction between Tantia Topee and Feroz Shah, Colonel Somerset decided to push on at once. He had marched forty miles on the 29th, and started at 3 A.M. on the morning of the 30th, but he hurried on none the less, and reached Kulcheepore at 5.30 P.M. At midnight (12.5 A.M. 31st December) he started again and marched on without a rest, except of an hour and a half to feed the horses, until 6.15 P.M., when he reached Satul after a forty-mile march. The rebels were now reported to be seven miles ahead, and it was determined, somewhat unfortunately, to march up to their encampment at once. As the British approached they were fired on by a rebel picquet; so that they could then do nothing more than lie down and wait till daylight. A small picquet of infantry, who had been riding on camels at the head of the column, was posted by the staff officer, and the Seventeenth then lay down on the ground, with their bridles in their hands. In a few moments every man was sound asleep. The staff-officer, waking an hour before daylight, found the bivouac like a camp of the dead—every soul so exhausted as to be overcome with sleep. The force was awakened without noise, and just at daylight the advance was resumed, but too late to overtake the rebels, who had moved off some time before. The British column, disregarding some dismounted soldiers and followers in the rebel camp, pushed on with all haste. The only track was of the worst possible description, and was necessarily allotted to the artillery, two troops of the Seventeenth trotting along, one on each flank of the guns, over the open. After thus traversing some seven miles, in the course of which

30th Dec.

31st Dec.

1858. the camels were left far in rear, the column came upon a village. The ground on each side thereof became impassable, so that the cavalry was compelled to bend outwards; and thus it came about that the guns, without escort, were actually the first to pass through a village with high walls, and with only just sufficient roadway to enable the guns to move. Fortunately the rebels made no effort to defend it; and it was only on debouching from the village that the gunners found, five hundred yards before them, three or four thousand rebel cavalry drawn up in line. Brigadier Somerset quietly turned to Major Paget, who commanded the half battery, and said "Gallop out towards them"; and so with the word "Leading gun, gallop," the formation of the British line began. The other guns then followed, and a staff officer galloped back to hurry forward the camel corps. Meanwhile the rebel cavalry advanced at a walk, one of their leaders on a gray horse endeavouring with all his might to induce his men to charge the guns. But the guns had unlimbered, and their very first shot swept away the gray horse. Some few rebels dismounted to pick up their chief, and the remainder of the force moved away to the British left. Then up came half a dozen of the 92nd on their camels; and then from each side of the village appeared the two troops of the Seventeenth. They numbered between eighty and ninety men all told, and came on in rank entire with lances at the "carry"—two small slender lines of pennons four hundred yards apart. "It was a pretty sight," says one who was there, "and the odds (4000 to 90) were so great that it became exciting also." Straight onward they galloped; and then suddenly the pennons swept forward like a flash of light, every lance came down to the "engage," and the Seventeenth with a yell dashed on to the charge. The rebels slackened pace, halted, and, before the lances had reached them, broke and fled; and the Seventeenth, plunging headlong among them, was swallowed up in the huge mass, and fairly vanished out of sight. Presently they appeared again, every lance still busy, and for seven miles the chase and the slaughter continued till men and horses could do no more.

Thus did the one squadron, so far unengaged, of the Seventeenth obtain its opportunity at last and take brilliant advantage thereof. A single man of the Seventeenth, wounded, summed up in himself the casualties of the whole column; but every soul was fairly worn out. Before the rebels were overtaken at Barode (for by this name the action is known), Somerset's column had marched a hundred and forty-seven miles without a halt except to feed the horses: the last fifty-two miles had been covered in thirty hours. The action with its pursuit of twelve miles made, with the return to camp, twenty-four miles more. All baggage and European supplies were left hopelessly in the rear: the nights were bitterly cold; and to bring discomfort to a climax, rain fell heavily for three days and three nights. Yet no one complained. On the morning after Barode men and horses were so numbed and stiff through cold and rain that they could hardly rise from the mud in which, through sheer fatigue, they had slept; and when after a few hours' painful march the sun at last broke through the clouds, the men gave him three cheers.

1859.
1st Jan.

But to Tantia, Barode was a mortal blow. The pursuing columns were now, so to speak, running for blood. General Michel shortly after the action formed a column wherein the whole of the Seventeenth was united, and pressed the chase with greater rapidity than ever, covering fifty-four miles and forty miles in two marches, and two hundred and fifty-six miles in eight days. On the 16th January, Tantia, flying northward, was caught and defeated by Brigadier Showers at Dewassa; on the 21st he was again caught and beaten by Colonel Holmes at Sikur. The Rao Sahib now abandoned Tantia in a rage, and Feroz Shah deserted him likewise. The former fled southward and was overtaken and defeated by Brigadier Honner's column near Koshani on the 10th February. On the 13th Brigadier Somerset took up the chase with three and a half squadrons of the Seventeenth in his column, and achieved a march which threw even his previous efforts into the shade. In six days and a half the Seventeenth

1859. covered no less than two hundred and thirty miles; they had their enemy dead-beat before them, and they knew it. Ghastly tokens met them on the march—hoof-tracks filled with blood, helpless innocent horses with their feet worn down to the quick, and, at the last, three hundred rebels who gave themselves up without a blow, being literally unable to run away any farther. The leaders alone escaped; but from that time the Rao Sahib's following ceased to exist; and he himself fled into the Banswarra jungle to be heard of no more. Tantia Topee, deserted, and since Sikur almost alone, hid in the Paron jungle until April, when he was betrayed by Rajah Man Singh to the English. He was tried by court-martial and hanged.

So ended this extraordinary chase, whereby the dying embers of the Mutiny were finally trampled out. In following the track of Tantia on the map, in and out and round about Malwa, one is reminded of nothing so much as the hunting of a rat in a barn. Though unendowed with the qualities that win success in a pitched battle, the man possessed a positive genius for guerilla warfare; and as he carried neither tents nor supplies, but satisfied his army's wants by the simple process of looting and stealing, he enjoyed always an advantage over his pursuers. His methods, in fact, differed little from those of the Pindaris, with whom the Seventeenth had to do in 1816-19; and but for the treachery of Rajah Man Singh he might have disappeared for ever into the jungle like his comrades the Rao Sahib and Feroz Shah, or met his fate at the jaws of a tiger like the Pindari chief Cheettoo.

Of the part played by the Seventeenth Lancers much has already been said in the course of the narrative. It now remains to add a few details which, lest the thread of the story should be unduly broken, have been reserved to the last.

First, we must note that in this campaign the Seventeenth wore its English clothing: blue tunic, overalls strapped with cloth, and forage cap protected by a white curtain, this last being preferred to the white-covered lance cap.

The bulk of the active work, as has been seen, fell upon Sir

William Gordon's squadron. When, after six months' hard work, Sir William rejoined the headquarters of the regiment, General Michel sent Colonel Benson the following letter :— 1859.

CAMP, MHOW, HEADQUARTERS, M.D.A.,
1st *December* 1858.

SIR,—I am directed by the Major-General to state that as the Seventeenth Lancers are again proceeding to take the field, he is desirous to express his strong approbation of the conduct of the squadron commanded by Sir William Gordon, which alone has accompanied the Mhow column through the whole of the late operations in the field.

2. Notwithstanding the most severe service in the worst weather, this squadron, owing to the unremitting attention of Sir W. Gordon, is almost as efficient as on the day when it left Mhow.

3. The Major-General has remarked that this officer's care was extended to the comfort of his men, the care of baggage animals, and even to the well-being of camp followers.

4. His leading in the field was as gallant as was his unremitting zeal; and in gallantry his officers and men emulated his example.

5. The Major-General, during the short time he has had under his personal observation the headquarters of your corps, has remarked with great pleasure that the general system of the regiment is one which must lead to efficiency; but this squadron has come so repeatedly under his observation in action and otherwise, that he cannot let it depart without specially recording his observation of its merits.

6. The Major-General directs that this letter may be read on parade of your regiment.—I have, &c.,

J. H. CHAPMAN, Capt., A.A.G., Malwa Division.

The most notable statement in this letter will be admitted to be that of the second paragraph :—

After the most severe service in the worst weather, this squadron, owing to the unremitting attention of Sir W. Gordon, is almost as efficient as on the day when it left Mhow.

This was no exaggeration. The squadron, for all its hard work, literally brought back every horse with which it had started fit for duty, excepting only those that had been killed or wounded in action; surely a performance of which any officer might well be proud. The troop-horses, it may be added, were mostly Arabs,

1859. and stood the work, by Sir William Gordon's testimony, remarkably well; and it is worth noting that in the supreme trial of two hundred and thirty miles in six days, several "walers" dropped dead under their riders, one or two Cape horses gave out, but no Arab was ever off his feed. We have already seen how Sir William Gordon took care of his horses, and we may now, by his kindness, catch a glimpse of his method of providing for those of whom he was even more careful—his men.

He writes as follows:—

> As a rule we had not much difficulty in getting supplies for men and horses, but occasionally had to resort to force. I remember on one occasion marching into a town called Samrood at 7 A.M. The head-man of the town kissed my feet in the saddle and promised that I should have all supplies at once. I thanked him, but as no supplies came I sent Evelyn Wood into the town with six men about 11 o'clock. They found abundance of everything required for men and horses, but no preparations to let us have what we wanted. So I ordered the head-man three dozen; after which he could not do enough for me, and supplies were plentiful. All was of course paid for; and the occurrence was reported by me to the authorities.

Let us not omit to add that the officer who took such care of his men and horses was himself a perfect horseman, having won the Regimental Challenge Cup within a few months of joining as a cornet; that, as we have seen, he fought the Russians at Balaclava till his head was almost cut to pieces; that at Mungrowlee he killed three men with his own hand, and throughout the Central Indian campaign frequently distinguished himself in personal combats; and that he has characteristically left the present writer to gather these latter details from any source except from himself.

Lastly, it must be remarked that this was the second if not the third campaign of its kind wherein the Seventeenth had been engaged. We saw it within twenty years of its foundation scouring the Carolinas and Virginia under Tarleton and Cornwallis, covering on one occasion one hundred and five miles in fifty-four hours, and traversing by constant forced marches a total distance of fifteen hundred miles. We found it next in Malwa in 1818 chasing the Pindaris; once making a forced march of thirty miles, and cutting Cheettoo's bandits to pieces at the end. Finally, forty years later, we follow it to this same Malwa through the mazy pursuit of Tantia Topee. In all three cases these incessant forced marches were accompanied by every hardship that could be inflicted by climate, privation, and fatigue; and whether we follow the Seventeenth in long-skirted scarlet and black helmet under the blazing sun of South Carolina and the drenching rain of the Alleghany slopes; or first in French gray jacket and white shako, and next in blue tunic and pugareed forage cap, through the burning days and bitter nights of the Malwa—in all three cases the story is the same. General Michel in 1858, no less than Lord Cornwallis in 1782, bears eloquent witness to the cheerful spirit and unconquerable patience with which these hardships were endured. Nor does the parallel hold less good of the action at the close of the march. It was when worn out with marching that a troop of the Seventeenth stood alone, after all others had given way, and cut its way through twenty times its number at Cowpens; it was when worn out with marching that a squadron of the Seventeenth charged and dispersed forty times its number at Barode.

1859.

CHAPTER XIV

PEACE SERVICE IN INDIA AND ENGLAND, 1859-1879

1859. FOR some time after the execution of Tantia the Seventeenth was kept marching about from day to day; and it was not until the 13th May that it finally went into quarters at Morar (Gwalior), detaching one squadron under Captain Taylor to Jhansi. In both places the regiment suffered severely from sickness, and lost many officers and men—the result of the climate, bad accommodation, and the reaction after the campaign. On the 10th January 1860 it was ordered to Secunderabad, and proceeded thither by rapid marches under command of Major White. On the way it lost thirty-eight more men of cholera and other diseases, among them Veigh, the butcher of the Balaclava charge, whose end was decidedly tragic. The deaths on the march, of course, entailed the digging of graves for the dead, in which work Veigh, who was a strong man and a thirsty soul, always glad to earn a few extra rupees, was particularly zealous. One day when his task of grave-digging was complete he was suddenly struck down by cholera, and in a few hours was buried in the grave which he had made for another. It was his final distinction to have dug his own grave.

1860.

1860-64. The regiment now remained at Secunderabad for five years. There is little to be chronicled of this period except one or two small matters of dress. In April 1860 the peaks on the forage caps were discontinued, and in 1861 the regiment, for the first time in its life, was equipped with white helmets. These were made of leather, covered with white cloth, without plume or spike, and were the work of a saddler sergeant who had come to the regiment from the 12th Lancers. 1864.

On the 14th December 1864 the Seventeenth left Secunderabad, and after sixteen days' march on foot arrived at Sholapore, whence it travelled by rail to Poona, and, after halting there till the 20th January 1865, reached Bombay, and embarked for England on the *Agamemnon* on the 21st. During the eight years of its service in India it was recruited at various times to a total number of 48 officers and 404 men. Its losses from climatic causes and disease, through death and invaliding, amounted to 38 officers and 373 men, while 122 more men were left behind as volunteers to serve with other regiments in India. 1865.

In April the regiment landed at Tilbury, and on the 6th May marched to Colchester, where it was inspected in October by the Commander-in-Chief, its sometime Colonel. Colonel White, the Commanding Officer, was now the only officer remaining who had ridden through the action at Balaclava, Sir William Gordon having retired in 1864. In the following year Colonel White retired, and was succeeded by Colonel Drury Lowe, a name that will live long in the regiment. It was in this same year 1866, the year of the Austro-Prussian war, that the Seventeenth were first quartered at Aldershot. 1866.

The year 1867 brings another name well known in the regiment on to the list of officers, this time not at the head of all, but at the foot of the cornets, that, namely, of John Brown, who held the adjutantcy from this time until 1878. Lieutenant-Colonel Brown (to give him his present rank) joined the Seventeenth as a band-boy in 1848. He rode the Balaclava charge as a trumpeter, and was brought to the ground close to the Russian battery, his horse's off hind leg being carried away by a cannon shot, and his own thigh pierced by a rifle bullet. After several weeks in hospital he rejoined the regiment in the Crimea, and when the Seventeenth went out to Central India dropped the trumpet for the lance. He was one of Major White's squadron at Barode, and from that time rose rapidly until he received his commission in 1867. For the 1867.

1867. present we need say no more than that he was Adjutant during Colonel Drury Lowe's command of the regiment.

In August 1867 the regiment was quartered at Shorncliffe 1868. and Brighton, where it remained until May 1868, when, after two months' stay at Woolwich, it was moved in August to Hounslow 1869. and Hampton Court. In the following year an experiment was tried which proved most successful, and has now been finally adopted, viz. the "squadron organisation." The squadron became the unit, and the word Troop was abolished—abolished, that is to say, in hope rather than in deed; for words which have the sanction of two centuries of use are not so easily expunged. When troops of cavalry first came into existence in England they were at least sixty men strong; when they were first organised by Statute they were one hundred men strong. Squadrons, again, were not compounds, but fractions of troops. Be that as it may, however, the old word Troop was for the time abolished, though not for long, and that of Squadron took its place. The establishment of cornets was, therefore, reduced by four; four troop sergeant-majors became squadron quartermaster-sergeants; four farriers were reduced and four shoeing-smiths added; and an additional sergeant (fencing instructor) was also added to the establishment. Simultaneously eight corporals and twenty-three privates were reduced, bringing down the total strength from 588 to 553, while the number of horses (a more serious matter) sank from 363 to 344.

In 1869 also the white plume, which had been adopted in 1857, was done away with, and a black plume issued in its stead. The original plume of the regiment, as we have seen, was scarlet and white, but was arbitrarily altered, for all Lancer regiments alike, by King William IV., to black. The old mourning lace, adopted by John Hale, having been long since abandoned, the black plume might seem to be a means of prolonging its memory; but the prejudice of the regiment ran in favour of white (scarlet and white being apparently out of date), and after a year or two the white plume was restored.

In July of the same year the regiment marched to Edinburgh and Hamilton, and remained in Scotland for ten months. This was its first visit to North Britain since 1760, when Colonel John Hale himself was in command. In 1870, as in 1764, the regiment moved from Scotland to Ireland—history thus repeating itself (if any one took notice of it) with commendable accuracy.

1869.

1870.

On the 15th August 1870 the establishment of the regiment was increased—the men from 457 to 540, the horses from 300 to 350. For France and Germany just then were flying at each other's throats, and even while the order was a-signing, were fighting the four days' battle (August 14-18) around Metz. As the outcome of this war, we shall have shortly to mention a number of sweeping reforms in the army. Meanwhile let us note that the first change of 1870, ordered before the war (1st April), was a retrograde step—a reversion to the old troop organisation. A step further back would have retained the name of a troop with the strength of a squadron, as in the days of the Ironsides. But the Army knows little of its own history.

With 1871 we enter on the first series of reforms, or let us call them changes, accomplished under the influence of the war of 1870.

1871.

First, the establishment of the regiment was fixed permanently at eight troops, after vacillating for more than a century between the minimum of six troops and the maximum of ten. Here, let us note, is a final break with the traditions of the great Civil War, when the six-troop organisation (each troop being 100 men strong) was first founded. Strictly speaking, the system of 1645 continued for some years later in the British regiments quartered in India; the Indian establishment consisting of six troops, while the other two formed a depôt in England; but this failing has now been remedied, and the old order is therefore wholly extinct.

Next, by Royal Warrant, the Purchase and Sale of Commissions in the Army were abolished. The system had existed for more than three hundred years, and had been threatened as far back as 1766.

1871. Next the "short service system"—six years' service with the colours and six in the reserve—was introduced; and thereby the old British soldier of history was, for good or ill, extinguished. The Seventeenth felt the change little before 1876; and the British public hardly found it out before 1879. It may be worth while to note that both short service[1] and the territorial system were first suggested just about a century before they were introduced.

Lastly, on the 1st November the historic rank of Cornet was abolished. *Corneta* or *cornette* signifies the horn-shaped troop standard which (like the ensign in the infantry) gave its name alike to the officer who carried it and to the troop that served under it. The rank is gone and all its historic associations with it; and a generation is arising which will need to resort to a dictionary if it would understand what Walpole meant when he called Pitt "that terrible cornet of horse." It is amusing to note that since the expurgation of the word Cornet no abiding name has been found for the rank of a junior subaltern of cavalry. Sub-lieutenants there have been and second lieutenants, sometimes both and sometimes neither, but nothing of permanence.

1872. The following year witnessed the death of another venerable institution, namely, of the "churns" carried by farriers. The name transports us to the days when farriers alone of cavalry men were dressed in blue and were armed with axes. The reintroduction of knee-boots, after an exile of sixty years, also revived, though in a different fashion, the memory of early days.

1873. The year 1873 likewise brought with it a reversion to primitive times in the shape of an order that greater attention should be paid to dismounted duty, the cavalry being now armed with the Snider carbine. This did not immediately affect the Seventeenth, which as yet possessed no carbines, but it was destined to do so before long. 1875. Two years later came another reform, this time in the matter of drill. The old system of

[1] The first hint of a short service system was given by a Frenchman, and presented, by translation, to England in 1590.

standing pivots, or as it was called the "pivot system," was abolished, and the "Evolutions" of 1759 lost their influence on cavalry drill for ever. 1875.

While all these changes were going forward the Seventeenth was quartered in Ireland, whither reform after reform pursued it across St. George's Channel. Being in Ireland it was, of course, called in to aid the civil power (Mallow election, 1872) but was spared the trouble of dealing with any disturbance. In 1876 it was brought over to England for mobilisation with the 5th Army Corps. Having called attention to the disavowal or attempted disavowal of the words Troop and Cornet, one cannot do less than emphasise the introduction of the comparatively strange terms, Mobilisation and Army Corps, which here confront the regiment for the first time. The Seventeenth was encamped on Pointing-down Downs in Somerset for a few weeks, and was reviewed with the 5th Army Corps on the 22nd July. As it is unlikely that the Seventeenth Lancers will ever again form part of a 5th Army Corps (for it is not often that England is so rich in army-corps) it seems well to record so unique an experience in a not uneventful career. 1876.

In this same year the Lancers' tunic was embellished with a plastron of the colour of the regimental facings,—a change which made the dress of the Seventeenth, by general admission, the smartest in the Army. The plastron being an essential feature in the uniform of the German Uhlan, is presumably imitated from Napoleon's Polish Lancers. No one will quarrel with so smart a dress; but it is nevertheless a little curious that the whole world should go to Poland for its Lancer fashions. The lance may be called the oldest of cavalry weapons, at least it can demonstrably be traced back beyond the days of Alexander the Great; and its present vogue is simply a return, and a late return, to an old favourite. Its reputation as the queen of cavalry weapons is not one century, but many centuries old; and though it was for a time driven out of the field by firearms, it may be said never to have wanted champions. I have found the lance advocated, for instance, by a

1876. French military writer in 1748, and by an English colonel, Dalrymple, in 1761. In 1590 the best authorities swore by it.

In 1876, likewise, came two more changes—the one temporary and the other permanent. The first was the issue of six carbines to every troop, a sign of a further change to come. The second was the appointment of the Duke of Cambridge to be Colonel-in-Chief of the regiment, which from henceforth is designated the "Duke of Cambridge's Own." In the early days of the Army it was customary on all occasions to insert the colonel's name after the regimental number; and thus it has been easy to identify the 18th (Hale's) Light Dragoons of 1759 with the present Seventeenth Lancers. The only colonels whose names enjoyed the distinction in the Seventeenth were Hale, Preston, and Gage. The Duke's name is now permanently bound with that of the regiment, a connection whereof, we trust, he will ever have good reason to feel proud.

1877. After staying at Aldershot until August 1877, the Seventeenth marched north to Leeds and Preston. After some service in aid of the civil power, which brought it at Clitheroe in collision with a mob of cotton operatives on strike, it returned to Aldershot in 1878. July 1878. A month later Colonel Drury Lowe retired, and was succeeded by Colonel Gonne. The Adjutant, Lieutenant John Brown, also resigned, but remained with the regiment as paymaster with the rank of captain.

In 1878 a change was made in the armament of the Seventeenth which takes us back to the earliest days of the British army. Martini-Henry carbines were issued, and pistols returned into store. Carbines, of course, were no new thing in the regiment, though they had been unknown therein since they were withdrawn (weapons very different from the Martini) in 1823. The bound from the old flint-lock to the Martini is remarkable; but the abolition of the pistol is even more noteworthy, for the pistol was a direct survival from the days of the Ironsides. Quite unconsciously the five regiments of Lancers carried the armament of Cromwell's troopers into the forty-first year of Queen Victoria.

As a weapon the pistol had long been regarded as of no account: it was a muzzle loader to the last, and as but ten rounds annually were allowed to each man for practice therewith, it was hardly taken seriously as a weapon at all. Still the abandonment of the pistol, as a point of historical interest, deserves at least so much notice. Sergeant-majors, and trumpeters were now provided with revolvers, a change which was fated to have serious influence on the careers of two officers of the regiment. 1878.

This year saw England committed to two wars, in Afghanistan and in Zululand. It must now be told how the Seventeenth Lancers played a part in both of them.

CHAPTER XV

THE ZULU WAR—PEACE SERVICE IN INDIA AND AT HOME, 1879-1894

1879. At the beginning of February England was shocked by the intelligence that one of Lord Chelmsford's columns, consisting of the 24th Regiment, had been surprised and annihilated by the Zulus at Isandlhwana (22nd January). The Seventeenth Lancers was at once warned to proceed on active service in South Africa, and the regiment was augmented by the transfer of sixty-five men and horses from the 5th and 16th Lancers. In the short interval between the warning and the embarkation the Commanding Officer, Colonel Gonne, was accidentally shot while superintending the practice of the non-commissioned officers with the newly issued revolver, and so severely wounded as to be unable to proceed on active service. Accordingly, on the 22nd February, Colonel Drury Lowe was gazetted as supernumerary Lieutenant-Colonel, and reassumed command of the regiment, his return being joyfully welcomed by all ranks, without exception, from the second in command downwards. On the same day the regiment was inspected by the Colonel-in-Chief at Hounslow, and two days later one wing, under the command of Major Boulderson, embarked on board the hired transport *France* at Victoria Docks; headquarters and the other wing embarking on board the *England* at Southampton on the 25th. A depôt of 121 men with 30 horses was left under the command of Captain Benson at Hounslow.

10th Feb.

24th Feb.

The strength of the regiment, as embarked, was as follows:—

	Field Officer.	Captains.	Subalterns.	Staff.	Total.	Rank and File.	Horses.		
							Officers.	Troopers.	Total.
Headquarter wing —*England*	1	4	7	4	16	302	25	238	263
Left wing—*France*	1	3	9	1	14	238	21	238	259
Totals	2	7	16	5	30	540	46	476	522

Both ships arrived at St. Vincent, Cape de Verdes, on the 7th March to coal; but owing to the great number of transports assembled at the same place for the same purpose, the *England* did not leave until the 12th, nor the *France* until the 14th. Both ships were detained again at Table Bay for a few days to coal, and arrived at Port Durban, the *England* on the 6th, and the *France* on the 11th April; five horses dead on the former, and six on the latter ship, were the casualties for the voyage. By the 14th both wings were disembarked, and the regiment then encamped for a day or two at Cator's Manor, near Durban—the right wing, under Colonel Drury Lowe, finally marching on the 17th April to Landman's Drift, and the left wing, under Major Boulderson, on the 21st April to Dundee.

The entire regiment shortly after marched up to Rorke's Drift together with the King's Dragoon Guards, the whole being under the command of Major-General Marshall. On the 21st May it visited the battlefield of Isandlhwana, buried most of the dead bodies, and brought back some of the abandoned waggons to Rorke's Drift. On the 23rd it joined the 2nd Division under Major-General Newdegate at Landman's Drift, on the 28th it marched with it to Koppie Allein on the Blood River, and at last on the 1st June crossed that river and entered Zululand.

On the 5th June the regiment came in contact with the Zulus for the first time at Erzungayan Hill. In a trifling skirmish which ensued the Adjutant, Lieutenant Frith, was shot dead by the

1879. 7th June.

Colonel's side. Two days later the division reached the Upoko River. A squadron of the Seventeenth was now detached to do duty at Fort Marshall, one of the posts constructed to guard the line of communication. The remainder moved up with division towards Ulundi, the kraal of the Zulu king. It was employed in the usual reconnaissance and outpost duties, varied by an occasional skirmish with the Zulus, but was never able to come to close quarters with the enemy. It was not employed, nor was any part of the strong force of cavalry available for the service, in a rapid advance upon Ulundi, as had been expected and hoped.

On the 2nd July the second division and flying column encamped on the south bank of the White Umvolosi River, about five miles from Ulundi, and on the 4th crossed the river and advanced against the kraal. The three squadrons of the Seventeenth formed the rear-guard; but no opportunity occurred of attacking the enemy on the march. The column was now rapidly enveloped by the Zulus in great force, and the cavalry was ordered to withdraw within the hollow square into which the infantry was formed. The Zulu attack began at 8.50 A.M., and was maintained for three-quarters of an hour within a hundred yards of a murderous artillery and rifle fire. During this time the Seventeenth stood to their horses under a heavy cross-fire, and suffered some casualties, Lieutenant Jenkins, among the officers, being shot in the jaw. About 9.30 the Zulus showed signs of wavering, and the Seventeenth was ordered out of the square to attack. As they rode out Captain Edgell was shot dead at the head of his squadron, and his troop farrier was killed at the same instant. Once clear of the square the regiment formed in echelon of wings, rank entire, covering over three hundred yards of front, and charged. It was met by a hot fire in front and flank from the Zulus, who were concealed in long grass in a donga; but charging right through them the Seventeenth scattered them in every direction, and then taking up the pursuit hunted them with great execution for nearly two miles. The horses were fresh, and there was no escape from the lances, which the enemy now encountered for the first time. The Zulu army was not only defeated but dispersed by this pursuit, and never appeared in the field again. The casualties of the Seventeenth on this day were, one officer (Captain Wyatt Edgell) and two men killed, three officers, viz. Colonel Drury Lowe, Lieutenant James, Scots Greys, attached to the Seventeenth, Lieutenant and acting Adjutant Jenkins, and five men wounded; the two first-named officers slightly, and the third severely. Also 26 horses were killed and wounded. The regiment was highly complimented, both verbally and in orders, by the General for its conduct at Ulundi. The only matter worthy of note in this short Zulu campaign is the heavy loss suffered by the Seventeenth in officers as compared with men; and this through pure chance, for all ranks were equally exposed.

1879.

The regiment began the return march on the day after the battle, with the 2nd Division, and arrived at the Upoko River on the 15th July. On the 26th it was ordered to march to Koppie Allein, to give over its horses to the King's Dragoon Guards, and to proceed dismounted to Pinetown, where it arrived on the 21st August. It was reduced a month later to six troops for Indian service; and 198 men then proceeded direct to England under Lieutenant W. Kevill-Davies. On the 1st October Colonel Drury Lowe for the second time took leave of the regiment; and Major Boulderson took command. The regiment then embarked for India; the left wing under Captain Cook sailing on board H.M.S. *Serapis* on 8th October, the right wing under Major Boulderson on board H.M.S. *Crocodile* on the 20th, and arriving at Bombay on the 28th October and 10th November respectively. The regiment was quartered at Mhow, the point from which it had started on the chase of Tantia Topee, twenty-one years before; the headquarters and the right wing arriving there on the 1st, and the left wing on the 14th November. Finally, on the 4th December Lieutenant-Colonel Gonne, who had recovered from his wound, arrived from England and took over the command. He was the only officer remaining in the regiment who had served with it in Central India in 1858-59.

The Seventeenth had not been long in India before a request

1880. came from General Phayre that the regiment might be sent up to join his force on active service in Afghanistan,—a request which, unfortunately, could not be complied with, owing to the defective state of the saddlery which was taken over in India. In July, however, twenty non-commissioned officers and men were sent up to do duty with the Transport on the Quetta-Candahar route. In this, as in all cases in the history of the regiment when small parties of men have been detached for particular duty, one and all did extremely well, and were complimented on the excellence of their work in an order published by the Commander-in-Chief of the Bombay Presidency. To make the parallel complete, two of these twenty now hold commissions—Major Forbes, the officer second in command of the King's Dragoon Guards, and Lieutenant Pilley, who remains with the Seventeenth as riding-master.

1881. In April of the following year Lieutenant-Colonel Gonne retired from the command, being appointed Military Attaché at St Petersburg; and in November Paymaster Captain John Brown took leave of the regiment with which he had been associated for five-and-thirty years. He and Major Berryman, the latter sometime the regimental Quartermaster, are the only two members of the Seventeenth who went through Balaclava, Central India, and South Africa.

The Seventeenth remained at Mhow until January 1884 without further incident worth the chronicling. Its old Colonel, General Drury Lowe, however, was meanwhile adding to his reputation in Egypt, where he commanded the cavalry division in the campaign of 1882. The pursuit of Arabi's army after the action of Tel-el-Kebir by the British cavalry, and the surrender of Cairo and of Arabi himself to General Drury Lowe, are matters of history. From the close of that campaign we must speak of him as Sir Drury Lowe, K.C.B.

1884. In February 1884 the Seventeenth Lancers relieved the 10th Hussars at Lucknow. In July Lieutenant-General Benson, who had commanded the regiment during the Central Indian campaign, became its Colonel. In December of the same year the regiment

A. Bassano, Photo. Walker & Boutall Ph. Sc.

Lieutenant General
Sir Drury C. Drury-Lowe, K.C.B.
Colonel 17th Lancers, 1892.

furnished a squadron to act as escort to the Commander-in-Chief in India, General Sir F. Roberts, at the camp of exercise in India. 1884.

The regiment remained at Lucknow until the expiration of its term of Indian service, embarking for England on H.M.S. *Serapis* on the 9th October 1890. One squadron was disembarked at Suez for duty with the army of occupation in Egypt, and was quartered at Abbasiyeh near Cairo. The remaining troops disembarked at Portsmouth on the 3rd of November. Of the non-commissioned officers and men who went out with the regiment to the Zulu War in 1879, just thirty returned with it in 1890; yet this was not due to death, for the Seventeenth lost but seventy men from disease during its last period of Indian service, an astonishing contrast to its former experiences in the times of the Pindari War and the Mutiny. For a year after its return the Seventeenth was quartered at Shorncliffe, where it was rejoined in November 1891 by the squadron that had been detached to Egypt, and then resumed the usual round of home service. The following year was marked by the successful introduction of the "squadron organisation," which had been already tried in 1869. 1890. 1891. 1892.

In January General Benson died, and the colonelcy of the regiment fell vacant. And as for the present we must close the history of the Seventeenth Lancers at this point, we cannot more fitly end it than with the name of General Benson's successor, the fifteenth and not the least Colonel of the regiment, Sir Drury Curzon Drury Lowe, K.C.B.

APPENDIX A

A LIST OF THE OFFICERS OF THE 17TH LIGHT DRAGOON LANCERS

NOTE.—The constant variation in the spelling of names in the earlier years of the regiment has made the preservation of uniformity in this respect a matter of great difficulty. I am still in doubt as to the correct method of spelling many names, and I can only plead that these doubts were shared by the owners of the names themselves.

1759

Lieutenant-Colonel.—John Hale
Major.—John Blaquière
Captains.—Franklin Kirby
Samuel Birch
Martin Basil
Edward Lascelles
John Burton
Samuel Townsend
Lieutenants.—Thomas Lee
William Green
Henry Wallop
Joseph Hall
Henry Cope
Yelverton Peyton
Cornets.—Robert Archdale
Henry Bishop
Joseph Stopford
Henry Crofton
Joseph Moxham
Daniel Brown
Adjutant.—Richard Westbury
Surgeon.—John Francis
Agent.—Mr. Calcraft, Channel Row, Westminster

1760-1761

Lieutenant-Colonel.—John Hale
Major.—John Blaquière
Captains.—Samuel Birch
Edward Lascelles
Charles Mawhood
John Burton
John Marriott
—— Baillie
Lieutenants.—Thomas Lea
William Green
Joseph Hall
Henry Wallop
Yelverton Peyton
N. Lane
Cornets.—Robert Archdale
Henry Bishop
Joseph Stopford
Henry Crofton
Joseph Moxham
Daniel Brown
George Birch
Francis Gwynne
James Poole
George Oliver

Cornet.—Samuel Burton
Adjutant.—Richard Westbury
Surgeon.—John Francis

1762

Lieut.-Colonel Commandant.—John Hale
Major.—John Blaquière
Captains.—Samuel Birch
Edward Lascelles
Charles Mawhood
John Burton
John Marriott
—— Baillie
Lieutenants.—Thomas Lea
William Green
Joseph Hall
Henry Wallop
Yelverton Peyton
N. Lane
Cornets.—Robert Archdale
Henry Bishop
Joseph Stopford
Henry Crofton
Joseph Moxham
Daniel Brown
George Birch
Francis Gwynne
James Poole
George Oliver
Samuel Burton
Richard Gwynne
Adjutant.—Richard Westbury
Surgeon.—John Francis

1763

Lieut.-Colonel Commandant.—John Hale
Major.—John Blaquière
Captains.—Samuel Birch
Charles Mawhood
John Marriott
Joseph Hall
Francis Lascelles
Captain.—Henry Bishop
Captain-Lieut.—Thomas Lea
Lieutenants.—Yelverton Peyton
N. Lane
Francis Jenison
Robert Archdale
Joseph Moxham
Cornets.—Henry Crofton
Daniel Brown
George Birch
Francis Gwynne
James Poole
George Oliver
Samuel Burton
Richard Gwynne
John Evans
Drury Wake
John Collings
Richard Parry
Adjutant.—Joseph Moxham
Surgeon.—John Francis

1764

Colonel.—John Hale
Lieut.-Colonel.—John Blaquière
Major.—Samuel Birch
Captains.—John Marriott
Joseph Hall
Henry Bishop
Thomas Lea
Captain-Lieut.—Yelverton Peyton
Lieutenants.—N. Lane
Robert Archdale
Joseph Moxham
Francis Gwynne
James Poole
Cornets.—Henry Crofton
Daniel Brown
George Evans
Harry Nettles
Benjamin Bunbury
Chaplain.—Thomas Ashcroft
Adjutant.—Joseph Moxham
Surgeon.—John Francis.

Appendix A

1765

Colonel.—John Hale
Lieut.-Colonel.—John Blaquière
Major.—Samuel Birch
Captains.—John Marriott
Joseph Hall
Henry Bishop
Thomas Lea
Captain-Lieut.—Yelverton Peyton
Lieutenants.—N. Lane
Robert Archdale
Joseph Moxham
Francis Gwynne
James Poole
Cornets.—Henry Crofton
Daniel Brown
George Evans
Harry Nettles
Benjamin Bunbury
Chaplain.—Thomas Ashcroft
Adjutant.—Joseph Moxham
Surgeon.—John Francis

1766

Colonel.—John Hale
Lieut.-Colonel.—John Blaquière
Major.—Samuel Birch
Captains.—Joseph Hall
Henry Bishop
Thomas Lea
Thomas S. Hall
Francis Gwynne
Captain-Lieut.—Robert Eyre
Lieutenants.—N. Lane
Robert Archdale
Joseph Moxham
James Poole
Harry Nettles
Cornets.—Benjamin Bunbury
Matthew Patteshall
Patrick Lynch
George Bennett
Hamlet Obins
Cornet.—John Francis
Chaplain.—Thomas Ashcroft
Adjutant.—Joseph Moxham
Surgeon.—William Waring

1767

Colonel.—John Hale
Lieut.-Colonel.—John Blaquière
Major.—Samuel Birch
Captains.—Henry Bishop
Thomas Lea
Francis Gwynne
James Poole
Francis Elliott
Captain-Lieut.—Robert Eyre
Lieutenants.—Nat. Lane
Robert Archdale
Joseph Moxham
Harry Nettles
Benjamin Bunbury
Cornets.—Matthew Patteshall
Hamlet Obins
John Francis
Martin Kerr
James Hussey
Frederick Metzer
Chaplain.—Thomas Ashcroft
Adjutant.—Joseph Moxham
Surgeon.—William Waring

1768

Colonel.—John Hale
Lieut.-Colonel.—John Blaquière
Major.—Samuel Birch
Captains.—Henry Bishop
Thomas Lea
Francis Gwynne
James Poole
Francis Elliott
Captain-Lieut.—Robert Eyre
Lieutenants.—N. Lane
Robert Archdale
Joseph Moxham
Harry Nettles

Lieutenant.—Benjamin Bunbury
Cornets.—Matthew Patteshall
Hamlet Obins
John Francis
Martin Kerr
James Hussey
Frederick Metzer
Chaplain.—Thomas Ashcroft
Adjutant.—John St. Clair
Surgeon.—William Waring

1769

Colonel.—John Hale
Lieut.-Colonel.—John Blaquière
Major.—Samuel Birch
Captains.—Henry Bishop
Thomas Lea
Francis Ed. Gwynne
James Poole
Arthur Blake
Captain-Lieut.—Robert Eyre
Lieutenants.—Robert Archdale
Joseph Moxham
Harry Nettles
Benjamin Bunbury
Matthew Patteshall
Cornets.—Hamlet Obins
John Francis
Martin Kerr
James Hussey
Frederick Metzer
Thomas Shadd
Chaplain.—James Adams
Adjutant.—John St. Clair
Surgeon.—Christopher Johnston

1770

Colonel.—John Hale
Lieut.-Colonel.—John Blaquière
Major.—Samuel Birch
Captains.—Henry Bishop
James Poole
C. Fortescue Garstin
Richard Carew
Captain.—Richard Gardiner
Captain-Lieut.—Joseph Moxham
Lieutenants.—Robert Archdale
Harry Nettles
Benjamin Bunbury
Matthew Patteshall
Hamlet Obins
Cornets.—John Francis
Martin Kerr
James Hussey
Frederick Metzer
Thomas Shadd
Thomas Whittaker
Chaplain.—James Adams
Adjutant.—John St. Clair
Surgeon.—Christopher Johnston

1771

Colonel.—George Preston
Lieut.-Colonel.—John Blaquière
Major.—Samuel Birch
Captains.—Henry Bishop
James Poole
C. Fortescue Garstin
T. Van Straubenzee
Vincent Corbet
Captain-Lieut.—Joseph Moxham
Lieutenants.—Robert Archdale
Harry Nettles
Benjamin Bunbury
Matthew Patteshall
Hamlet Obins
Cornets.—John Francis
Mark Kerr
James Hussey
Frederick Metzer
Thomas Whittaker
William Loftus
Chaplain.—James Adams
Adjutant.—John St. Clair
Surgeon.—Christopher Johnston

1772

Colonel.—George Preston

Lieut.-Colonel.—John Blaquière
Major.—Samuel Birch
Captains.—Henry Bishop
James Poole
C. Fortescue Garstin.
T. Van Straubenzee
Vincent Corbet
Captain-Lieut.—Joseph Moxham
Lieutenants.—Robert Archdale
Harry Nettles
Benjamin Bunbury
Matthew Patteshall
Hamlet Obins
Cornets.—John Francis
Mark Kerr
James Hussey
Frederick Metzer
Thomas Whittaker
William Loftus
Chaplain.—James Adams
Adjutant.—John St. Clair
Surgeon.—Christopher Johnston

1773

Colonel.—George Preston
Lieut.-Colonel.—John Blaquière
Major.—Samuel Birch
Captains.—Henry Bishop
C. Fortescue Garstin
T. Van Straubenzee
Richard Crewe
Joseph Moxham
Captain-Lieut.—Robert Archdale
Lieutenants.—Harry Nettles
Benjamin Bunbury
Matthew Patteshall
Hamlet Obins
John Francis
Cornets.—Mark Kerr
James Hussey
Frederick Metzer
Thomas Whittaker
William Loftus
John St. Clair
Chaplain.—Richard Griffith
Adjutant.—John St. Clair
Surgeon.—Christopher Johnston

1774

Colonel.—George Preston
Lieut.-Colonel.—John Blaquière
Major.—Henry Bishop
Captains.—C. F. Garstin
Richard Carew
T. Van Straubenzee
Joseph Moxham
Oliver Delancey
Captain-Lieut.—Robert Archdale
Lieutenants.—Henry Nettles
Benjamin Bunbury
Matthew Patteshall
H. Obins
John Francis
Mark Kerr
Cornets.—James Hussey
Frederick Metzer
Thomas Whittaker
William Loftus
John St. Clair
Chaplain.—Richard Griffith
Adjutant.—John St. Clair
Surgeon.—Christopher Johnston

1775

Colonel.—George Preston
Lieut.-Colonel.—John Blaquière
Major.—Henry Bishop
Captains.—C. F. Garstin
Richard Crewe
T. Van Straubenzee
Joseph Moxham
Oliver Delancey
Hon. F. Needham
Captain-Lieut.—Robert Archdale
Lieutenants.—Harry Nettles
Benjamin Bunbury
Matthew Patteshall
H. Obins

Lieutenants.—John Francis
Mark Kerr
Cornets.—James Hussey
Frederick Metzer
Thomas Whittaker
William Loftus
John St. Clair
Samuel Bagot
Thomas J. Cook
Chaplain.—Richard Griffith
Adjutant.—John St. Clair
Surgeon.—Christopher Johnston

1776

Colonel.—George Preston
Lieut.-Colonel.—John Blaquière
Major.—Henry Bishop
Captains.—C. F. Garstin
Richard Crewe
T. V. Straubenzee
Joseph Moxham
Oliver Delancey
Hon. F. Needham
Captain-Lieut.—Robert Archdale
Lieutenants.—Harry Nettles
Benjamin Bunbury
Matthew Patteshall
H. Obins
John Francis
Mark Kerr
Cornets.—James Hussey
Frederick Metzer
William Loftus
John St. Clair
Samuel Bagot
William St. Leger
David Ogilvy
David St. Clair
John Sloper
Peter Anderson
John Hamilton
Chaplain.—Richard Griffith
Adjutant.—John St. Clair
Surgeon.—Christopher Johnston

1777

Colonel.—George Preston
Lieut.-Colonel.—Samuel Birch
Major.—Richard Crewe
Captains.—Joseph Moxham
Oliver Delancey
Hon. F. Needham
Hon. Thomas Stanley
R. H. Elliston
Captain-Lieut.—Robert Archdale
Lieuts.—Harry Nettles
Matthew Patteshall
Mark Kerr
James Hussey
Geo., Visct. Deerhurst
Cornets.—Frederick Metzer
John St. Clair
Samuel Bagot
David Ogilvy
John Sloper
Peter Anderson
John Hamilton
Thomas Patterson
John Jones
Samuel Watts
William St. Leger
Chaplain.—Richard Griffith
Adjutant.—John St. Clair
Surgeon.—Christopher Johnston

1778

Colonel.—George Preston
Lieut.-Colonel.—Samuel Birch
Major.—Richard Crewe
Captains.—Joseph Moxham
Oliver Delancey
Hon. F. Needham
Hon. Thomas Stanley
R. H. Elliston
Captain-Lieut.—Robert Archdale
Lieutenants.—Harry Nettles
Matthew Patteshall
Mark Kerr

Lieutenants.—James Hussey
Geo., Visct. Deerhurst
Wm., Lord Cathcart
Cornets.—Frederick Metzer
John St. Clair
Samuel Bagot
David Ogilvy
John Sloper
John Hamilton
Thomas Patterson
John Jones
Samuel Watts
William St. Leger
Thomas Romain
T. Smith Bradshaw
Chaplain.—Richard Griffith
Adjutant.—John St. Clair
Surgeon.—Christopher Johnston

1779

Colonel.—George Preston
Lieut.-Colonel.—Samuel Birch
Major.—Oliver Delancey
Captains.—Hon. F. Needham
Wm. Lord Cathcart
Wm. Henry Talbot
(Two vacancies)
Captain-Lieut.—Robert Archdale
Lieutenants.—Harry Nettles
Matthew Patteshall
Mark Kerr
James Hussey
Samuel Bagot
Cornets.—William St. Leger
David Ogilvy
John Sloper
John Hamilton
John Jones
T. Smith Bradshaw
J. Stapleton
Thomas Patterson
Charles Searle
John St. Clair
J. Thos. Fonblanque
Chaplain.—Richard Griffith
Adjutant.—John St. Clair
Surgeon.—Christopher Johnston

1780

Colonel.—George Preston
Lieut.-Colonel.—Samuel Birch
Major.—Oliver Delancey
Captains.—Hon. F. Needham
Wm. Henry Talbot
Samuel Bagot
Captain-Lieut.—Robert Archdale
Lieutenants.—Harry Nettles
Matthew Patteshall
Mark Kerr
James Hussey
T. Smith Bradshaw
Cornets.—David Ogilvy
John Jones
J. Stapleton
Thomas Patterson
Charles Searle
John St. Clair
J. Thos. Fonblanque
Thomas Tucker
John Black
Chaplain.—John Beevor
Adjutant.—John Jones
Surgeon.—Christopher Johnston
Agents.—Cox, Muir & Co.

1781

Colonel.—George Preston
Lieut.-Colonel.—Samuel Birch
Major.—Oliver Delancey
Captains.—Robert Archdale
Wm. Henry Talbot
Samuel Bagot
T. Smith Bradshaw
Captain-Lieut.—John Stapleton
Lieutenants.—Harry Nettles
Matthew Patteshall
Mark Kerr
James Hussey

Lieutenant.—John Jones
Cornets.—Thomas Patterson
Charles Searle
John St. Clair
Thomas Tucker
John Black
David M'Culloch
Warren Delancey
Joseph White
Chaplain.—John Beevor
Adjutant.—John Jones
Surgeon.—Christopher Johnston
Agents.—Cox, Muir & Co.

1782

Colonels.—George Preston
Samuel Birch
Major.—Oliver Delancey
Captains.—Robert Archdale
Wm. Henry Talbot
Samuel Bagot
T. Smith Bradshaw
Captain-Lieut.—John Stapleton
Lieutenants.—Harry Nettles
Matthew Patteshall
Mark Kerr
James Hussey
John Jones
Cornets.—Thomas Patterson
Charles Searle
John St. Clair
Thomas Tucker
John Black
Warren Delancey
Joseph White
David MacCulloch
William Jephson
William Woodley
Chaplain.—John Beevor
Adjutant.—John Jones
Surgeon.—Christopher Johnston

1783

Colonel.—Hon. Thomas Gage
Lieut.-Colonel.—Samuel Birch
Major.—Oliver Delancey
Captains.—Robert Archdale
Samuel Bagot
T. Smith Bradshaw
John Stapleton
Captain.-Lieut.—Harry Nettles
Lieutenants.—Matthew Patteshall
Mark Kerr
James Hussey
John Jones
Henry G. Grey
Cornets.—John St. Clair
Thomas Tucker
John Black
Warren Delancey
William Jephson
Joseph White
William Woodley
George Birch
C. L. Wallace
Ralph Hamilton
Chaplain.—John Beevor
Adjutant.—John Jones
Surgeon.—Christopher Johnston
Agents.—Cox, Muir & Co.

1784

Colonel.—Hon. Thomas Gage
Lieut.-Colonel.—Samuel Birch
Major.—Oliver Delancey
Captains.—Robert Archdale
Samuel Bagot
John Stapleton
Captain-Lieut.—Harry Nettles
Lieutenants.—James Hussey
John Jones
Henry G. Grey
John Black
Cornets.—John St. Clair
William Jephson
Joseph White
Francis E. Lee
Chaplain.—John Beevor

Adjutant.—John Jones
Surgeon.—Christopher Johnston
Agents.—Cox, Muir & Co.

1785

Colonel.—Thomas, Earl of Lincoln
Lieut.-Colonel.—Samuel Birch
Major.—Oliver Delancey
Captains.—Robert Archdale
Samuel Bagot
John Stapleton
William St. Leger
Captain-Lieut.—Harry Nettles
Lieutenants.—John Jones
Henry G. Grey
John Black
Thomas Tucker
William Hatton
Cornets.—William Jephson
Joseph White
Evan Lloyd
Richard Odlum
R. F. Currie
Chaplain.—John Beevor
Adjutant.—John Jones
Surgeon.—Christopher Johnston

1786

Colonel.—Thomas, Earl of Lincoln
Lieut.-Colonel.—Samuel Birch
Major.—Oliver Delancey
Captains.—Robert Archdale
Samuel Bagot
John Stapleton
William St. Leger
Captain-Lieut.—Harry Nettles
Lieutenants.—John Jones
John Black
Thomas Tucker
William Hatton
Cornets.—William Jephson
Joseph White
Richard Odlum
Cornets.—R. F. Currie
William Wells
Francis E. Lee
Chaplain.—A. Greenfield
Adjutant.—John Jones
Surgeon.—Christopher Johnston
Agents.—Wybrants & Son, Dublin

1787

Colonel.—Thomas, Earl of Lincoln
Lieut.-Colonel.—Samuel Birch
Major.—Oliver Delancey
Captains.—Robert Archdale
Samuel Bagot
John Stapleton
William St. Leger
Captain-Lieut.—Harry Nettles
Lieutenants.—John Jones
John Black
Thomas Tucker
William Hatton
Cornets.—William Jephson
Joseph White
Evan Lloyd
Richard Odlum
Francis E. Lee
Samuel Stapleton
P. D. du Moulin
Chaplain.—A. Greenfield
Adjutant.—John Jones
Surgeon.—Christopher Johnston
Agents.—Wybrants & Son, Dublin

1788

Colonel.—Thomas, Earl of Lincoln
Lieut.-Colonel.—Samuel Birch
Major.—Oliver Delancey
Captains.—Robert Archdale
Samuel Bagot
John Stapleton
William St. Leger
Captain-Lieut.—Harry Nettles
Lieutenant.—John Jones

Lieutenants.—John Black
Thomas Tucker
Evan Lloyd
William Jephson
Cornets.—Joseph White
Richard Odlum
Francis E. Lee
Samuel Stapleton
P. D. du Moulin
Thomas Grey
Chaplain.—A. Greenfield
Adjutant.—John Jones
Surgeon.—Christopher Johnston
Agents.—Wybrants & Son, Dublin

1789

Colonel.—Thomas, Earl of Lincoln
Lieut.-Colonel.—Samuel Birch
Major.—Oliver Delancey
Captains.—Robert Archdale
Samuel Bagot
William St. Leger
George Pigott
Captain-Lieut.—Harry Nettles
Lieutenants.—John Jones
John Black
Evan Lloyd
William Jephson
Joseph White
Cornets.—Richard Odlum
Francis E. Lee
Samuel Stapleton
P. D. du Moulin
Thomas Grey
William S. Bacon
Chaplain.—Thomas Sneyd
Adjutant.—John Jones
Surgeon.—Christopher Johnston
Agents.—Wybrants & Son, Dublin

1790

Colonel.—Thomas, Earl of Lincoln
Lieut.-Colonel.—Samuel Birch
Major.—Oliver Delancey
Captains.—Robert Archdale
Samuel Bagot
George Pigott
Hon. John Hope
Captain-Lieut.—Harry Nettles
Lieutenants.—John Jones
John Black
Evan Lloyd
William Jephson
Richard Odlum
Cornets.—Frank E. Lee
Peter D. du Moulin
Thomas Grey
William S. Bacon
Christopher Johnston
Chaplain.—Thomas Sneyd
Adjutant.—John Jones
Surgeon.—Christopher Johnston
Agents.—Wybrants & Son, Dublin

1791

Colonel.—Thomas, Earl of Lincoln
Lieut.-Colonel.—Samuel Birch
Major.—Oliver Delancey
Captains.—Robert Archdale
George Pigott
Hon. John Hope
Captain-Lieut.—Harry Nettles
Lieutenants.—John Jones
John Black
Evan Lloyd
William Jephson
Richard Odlum
Cornets.—Francis E. Lee
Peter D. du Moulin
Thomas Grey
William S. Bacon
Christopher Johnston
Chaplain.—Thomas Sneyd
Adjutant.—John Gibson
Surgeon.—Christopher Johnston
Agents.—Wybrants & Son, Dublin

Appendix A

1792

Colonel.—Thomas, Earl of Lincoln
Lieut.-Colonel.—Samuel Birch
Major.—Oliver Delancey
Captains.—Robert Archdale
George Pigott
Hon. John Hope
Captain-Lieut.—Harry Nettles
Lieutenants.—John Jones
John Black
Evan Lloyd
William Jephson
Richard Odlum
Cornets.—Peter David du Moulin
William S. Bacon
Christopher Johnston
(3 vacancies)
Chaplain.—Thomas Sneyd
Adjutant.—John Gibson
Surgeon.—Christopher Johnston

1793

Colonel.—Thomas, Earl of Lincoln
Lieut.-Colonel.—Samuel Birch
Major.—Oliver Delancey
Captains.—George Pigott
Charles Maitland
John Jones
Captain-Lieut.—Harry Nettles
Lieutenants.—John Black
Evan Lloyd
William Jephson
Richard Odlum
William S. Bacon
Cornets.—Peter D. du Moulin
Christopher Johnston
William Richards
Oswald Werge
Leonard Shafto Orde
Theobald Butler
Chaplain.—Thomas Sneyd
Adjutant.—Edward Wilson
Surgeon.—Christopher Johnston

1794

Colonel.—Thomas, Duke of Newcastle
Lieut.-Colonel.—Samuel Birch
Major.—Oliver Delancey
Captains.—George Pigott
Charles Maitland
John Jones
Captain-Lieut.—Harry Nettles
Lieutenants.—John Black
Evan Lloyd
William Jephson
Richard Odlum
William S. Bacon
Cornets.—Christopher Johnston
William Richards
Oswald Werge
Theobald Butler
William L. Murray
Chaplain.—Thomas Sneyd
Adjutant.—John Mainwaring
Surgeon.—Christopher Johnston

1795

Colonel.—Thomas, Duke of Newcastle
Lieut.-Colonel.—Oliver Delancey
Major.—Harry Nettles
Captains.—Charles Maitland
John Jones
Evan Lloyd
Hon. John Creighton
John Black
William L. Murray
Captain-Lieut.—William Jephson
Lieutenants.—Richard Odlum
William S. Bacon
Christopher Johnston
William Richards
Oswald Werge
Thomas Butler
(2 vacancies)
Cornet.—Samuel Bristow

Cornets.—Richard Aylmer
Richard Garstin
John Jones
Edward Wilson
Richard Edwards
David Supple
(2 vacancies)
Chaplain.—Thomas Sneyd
Adjutant.—John Mainwaring
Surgeon.—Christopher Johnston

1796

Colonel.—Oliver Delancey
Lieut.-Colonel.—George Hardy
Majors.—Harry Nettles
Evan Lloyd
Captains.—John Black
William Jephson
Francis Gore
Robert Fletcher
Robert Lowe
James MacDonell
Capt.-Lieut.—Christopher Johnston
Lieutenants.—William Richards
Oswald Werge
Thomas Butler
Richard Aylmer
Richard Garstin
Edward Wilson
Richard Edwards
David Supple
Cornets.—John Mainwaring
James Byrne
John Gildea
Philip Teesdale
James Hellings
John Jones
Thomas Smithson
John Delancey
William Grey
John Willington
Chaplain.—Thomas Sneyd
Adjutant.—John Mainwaring
Surgeon.—John Robinson

1797

Colonel.—Oliver Delancey
Lieut.-Colonel.—Henry George Grey
Majors.—Evan Lloyd
William Jephson
Captains.—Francis Gore
Robert Fletcher
Robert Lowe
James MacDonell
Christopher Johnston
William H. Delancey
Captain-Lieut.—William Richards
Lieutenants.—Oswald Werge
Richard Aylmer
Richard Garstin
Edward Wilson
Richard Edwards
David Supple
John Mainwaring
James Byrne
Philip Teesdale
James Hellings
John Jones
John Delancey
Cornets.—Jon. Willington
John Jappie
Thomas Glegg
Thomas A. Cookson
Chaplain.—Thomas Sneyd
Adjutant.—John Mainwaring
Surgeon.—John Robinson

1798

Colonel.—Charles Delancey
Lieut.-Colonel.—H. G. Grey
Majors.—Evan Lloyd
William Jephson
Captains.—Francis Gore
Robert Fletcher
Robert Lowe
James MacDonell
Christopher Johnston
William H. Delancey
Captain-Lieut.—William Richards

Lieutenants.—Oswald Werge
Richard Aylmer
Richard Garstin
Edward Wilson
Richard Edwards
David Supple
John Mainwaring
Philip Teesdale
James Hellings
John Delancey
Peter Carey
J. Cocks
Vere L. Ward

Cornets.—Jon. Willington
John Werge
John Jappie
Thomas Ahmuty
John M. Winter
Thomas Cockerill
William Roycraft

Adjutant.—William Roycraft
Surgeon.—William Robinson
Asst.-Surgeon.—Thomas Thompson
Veterinary-Surgeon.—James Burt

1799

Colonel.—Oliver Delancey
Lieut.-Colonels.—H. G. Grey
Evan Lloyd
Majors.—William Jephson
Francis Gore
Captains.—Robert Lowe
James MacDonell
Christopher Johnston
William H. Delancey
William Richards
Robert Jones
Captain-Lieut.—Oswald Werge
Lieutenants.—Richard Aylmer
Richard Garstin
Edward Wilson
Richard Edwards
David Supple
John Mainwaring

Lieutenants.—Philip Teesdale
James Hellings
John Delancey
Peter Carey
J. Cocks
V. L. Ward
Jon. Willington
Cornets.—John Werge
John Jappie
Thomas Ahmuty
William Roycraft
Thomas Cockerill
William Ogden
John Laing
James O'Reilly
John Clarke
Adjutant.—William Roycraft
Surgeon.—William Robinson
Assistant-Surgeon.—Lewis Bowen
Veterinary-Surgeon.—James Burt
Paymaster.—James Byrne
Agents.—Cox & Company
1796.—Chaplain discontinued
1797.—Assistant-Surgeon appointed
1798.—Paymaster appointed
1799.—A second Lieut.-Colonel appointed

1800

Colonel.—Oliver Delancey
Lieut.-Colonels.—H. G. Grey
Evan Lloyd
Majors.—William Jephson
Francis Gore
Captains.—Robert Lowe
James MacDonell
Christopher Johnston
Oswald Werge
Richard Aylmer
John Daniell
Thomas Ellis
Thomas Gerrard
Captain-Lieut.—Edward Wilson
Lieutenant.—David Supple

Lieutenants.—John Mainwaring
Philip Teesdale
James Hellings
Peter Carey
Jon. Willington
R. K. Carden
John Werge
John Laing
John Delancey
P. K. Roche
Cornets.—John Jappie
Thomas Ahmuty
William Roycraft
Thomas Cockerill
Henry Harris
Joseph Hawtyn
George Lang
James Annesley
Edward Kelly
H. W. Thompson
Adjutant.—William Roycraft
Surgeon.—William Robinson
Assistant-Surgeon.—Lewis Bowen
Veterinary Surgeon.—James Burt
Paymaster.—James Byrne
Agents.—Cox & Company
(A second Assistant-Surgeon appointed)

1801

Colonel.—Oliver Delancey
Lieut.-Colonels.—H. G. Grey
Evan Lloyd
Majors.—William Jephson
Francis Gore
Captains.—James MacDonell
Robert Lowe
Christopher Johnston
Oswald Werge
Richard Aylmer
John Daniell
Thomas Ellis
Thomas Gerrard
Captain-Lieut.—Edward Wilson
Lieutenants.—David Supple
John Mainwaring
Philip Teesdale
James Hellings
Peter Carey
Jon. Willington
John Werge
John Laing
Wm. Ch. Jerningham
P. K. Roche
Cornets.—John Jappie
William Roycraft
Thomas Cockerill
Henry Harris
Joseph Hawtyn
George Lang
James Annesley
William J. Kent
W. B. Laird
Joseph Tyndale
Adjutant.—William Roycraft
Surgeon.—William Robinson
Asst.-Surgeons.—Samuel Tilt
Alexander Menzies
Veterinary Surgeon.—James Peers
Paymaster.—James Byrne
Agents.—Cox & Company

1802

Colonel.—Oliver Delancey
Lieut.-Colonels.—H. G. Grey
Evan Lloyd
Majors.—William Jephson
Francis Gore
Captains.—James MacDonell
Robert Lowe
Christopher Johnston
Oswald Werge
Richard Aylmer
John Daniell
Thomas Ellis
Thomas Gerrard
Captain-Lieut.—Edward Wilson
Lieutenant.—David Supple

Lieutenants.—John Mainwaring
Philip Teesdale
James Hellings
Jonathan Willington
John Werge
P. K. Roche
Wm. Ch. Jerningham
W. B. Laird
John Jappie
William Roycraft
Thomas Cockerill
Henry Harris
Joseph Hawtyn
Henry F. R. Soane
Richard Miller
James Annesley

Cornets.—William J. Kent
Joseph Tyndale
Montfort Westropp
William Brown
Edmund Safferey
—— Gledd
—— Brydges
De Lancey Barclay

(Staff as in previous year)

1803

Colonel.—Oliver Delancey

Lieut.-Colonels.—H. G. Grey
Evan Lloyd

Majors.—William Jephson
James MacDonell

Captains.—Robert Lowe
Christopher Johnston
Oswald Werge
Richard Aylmer
John Daniell

Captain-Lieut.—Edward Wilson

Lieutenants.—David Supple
John Mainwaring
Philip Teesdale
James Hellings
Jonathan Willington
P. K. Roche
Wm. Ch. Jerningham
W. Roycraft
De Lancey Barclay

Cornets.—Joseph Tyndale
Montfort Westropp
William Brown
Edmund Safferey
—— Gledd
Thomas Turner

Paymaster.—James Byrne

Adjutant.—William Roycraft

Surgeon.—William Robinson

Asst.-Surgeons.—Samuel Tilt
Alexander Menzies

Veterinary Surgeon.—James Peers

1804

Colonel.—Oliver Delancey

Lieut.-Colonels.—H. G. Grey
Evan Lloyd

Majors.—William Jephson
James MacDonell

Captains.—Robert Lowe
Christopher Johnston
Oswald Werge
Richard Aylmer
Edward Wilson
John Werge
W. B. Laird
David Supple

Lieutenants.—Philip Teesdale
James Hellings
Jonathan Willington
P. K. Roche
William Roycraft
De Lancey Barclay
Montfort Westropp
Edmund Safferey
Thomas Turner

Cornets.—William Brown
John Sharland Harris
J. R. L. Lloyd
William C. Faulkner
William D'Arcy

Cornet.—William Moray
Paymaster.—James Byrne
Adjutant.—William Roycraft
Surgeon.—James O'Connor
Assistant-Surgeon.—Samuel Tilt
Veterinary Surgeon.—James Peers

1805

Colonel.—Oliver Delancey
Lieut.-Colonels.—H. G. Grey
Evan Lloyd
Majors.—James MacDonell
Christopher Johnston
Captains.—Oswald Werge
Richard Aylmer
Edward Wilson
John Daniell
John Werge
W. B. Laird
David Supple
Philip Teesdale
James Hellings
P. K. Roche
Lieutenants.—Jonathan Willington
William Roycraft
De Lancey Barclay
Edmund Safferey
Thomas Turner
William Brown
Hon. John Jones
W. C. Faulkner
William D'Arcy
J. R. Lloyd
William Moray
Cornets.—Ralph Laurence
Robert D'Arcy
James Reid
Charles Johnson
William Abbs
(2 vacancies)
Paymaster.—James Byrne
Adjutant.—William Roycraft
Surgeon.—James Anderson
Assistant-Surgeon.—Samuel Tilt
Assistant-Surgeon.—John Hemphill
Vet. Surg.—Edward Coleman

1806

Colonel.—Oliver Delancey
Lieutenant-Colonels.—H. G. Grey
Evan Lloyd
Majors.—James MacDonell
Henry Loftus
Captains.—Oswald Werge
Edward Wilson
John Daniell
W. B. Laird
David Supple
Philip Teesdale
James Hellings
P. K. Roche
Francis D'Arcy Bacon
Archibald Ross
Lieutenants.—Jonathan Willington
William Roycraft
Edmund Safferey
William Brown
Hon. John Jones
W. C. Faulkner
William D'Arcy
J. R. L. Lloyd
Wm. Moray
Robert D'Arcy
Ralph Lawrenson
James Read
Henry Walker
John Burton
Frederick Willoe
Charles Johnson
Benjamin Adams
John Blake
Cornets.—James Delancey
John Lane
Edward Wrixon
Charles White
Bartholomew Thomas
Frederick Geale
Thomas Lahiff

Cornet.—James Butler
(Staff as in 1805)
Agents.—Messrs. Arnutt & Brough, Dublin

1807

Colonel.—Oliver Delancey
Lieut.-Colonels.—Hon. H. G. Grey
Evan Lloyd
Majors.—Henry Loftus
Lynch Cotton
Captains.—Oswald Werge
Edward Wilson
John Daniell
William B. Laird
David Supple
Philip Teesdale
James Hellings
P. K. Roche
F. D. Bacon
Archibald Ross
Lieutenants.—Jonathan Willington
William Roycraft
Edmund Safferey
William Brown
Hon. John Jones
William D'Arcy
Ralph Lawrenson
James Read
Henry Walker
John Burton
Frederick Willoe
Charles Johnson
Benjamin Adams
John Blake
James Delancey
Cornets.—John Lane
Edward Wrixon
Bartholomew Thomas
Frederick Geale
Thomas Lahiff
James Butler
G. W. R. Lewin
Paymaster.—James Byrne
Adjutant.—William Roycraft
Surgeon.—James Anderson
Assistant-Surgeons.—James Tilt
—— Howship
Vet. Surg.—Edward Coleman

1808

Colonel.—Oliver de Lancey
Lieut.-Colonels.—Hon. H. G. Grey
Evan Lloyd
Majors.—Henry Loftus
Lynch Cotton
Captains.—Oswald Werge
Edward Wilson
John Daniell
William B. Laird
David Supple
Philip Teesdale
P. K. Roche
Francis D. Bacon
Archibald Ross
Jonathan Willington
Lieutenants.—William Roycraft
Edmund Safferey
William Brown
Hon. John Jones
William D'Arcy
J. R. L. Lloyd
Robert D'Arcy
William Moray
James Read
Henry Walker
John Burton
Frederick Willoe
Charles Johnson
Benjamin Adams
John Blake
James de Lancey
John Lane
Cornets.—Edward Wrixon
Bartholomew Thomas
Frederick Geale
Thomas Lahiff
James Butler

Cornet.—G. W. R. Lewin
Paymaster.—(Vacant)
Adjutant.—William Roycraft
Surgeon.—James Anderson
Assistant-Surgeons.—Samuel Tilt
—— Howship
Vet. Surg.—Edward Coleman

1809

Colonel.—Oliver de Lancey
Lieut.-Colonels.—Hon. H. G. Grey
Evan Lloyd
Majors.—Henry Loftus
Lynch Cotton
Captains.—Oswald Werge
David Supple
Philip Teesdale
Jonathan Willington
James Grant
George John Sale
William Moray
Henry Yonge
Thomas Forster
Henry Walker
William Roycraft
Lieutenants.—Edmund Safferey
William Brown
Hon. John Jones
J. R. L. Lloyd
James Read
John Burton
Frederick Willoe
Charles Johnson
Benjamin Adams
Thomas Lahiff
Edward Wrixon
G. W. Wallace
John Brackenbury
H. E. Lynch
John D'Arcy
—— Johnson
William Gale
Cornets.—G. W. R. Lewin
James Tomkinson
Michael Ryan
Joseph Budden
William Henry Robinson
Paymaster.—Robert Harman
Adjutant.—William Gale
Surgeon.—William King
Assistant-Surgeons.—John White
David Christie
Vet. Surg.—Edward Coleman

1810

Colonel.—Oliver de Lancey
Lieut.-Colonels.—Hon. H. G. Grey
Evan Lloyd
Majors.—Oswald Werge
Charles Morland
Captains.—David Supple
Philip Teesdale
Jonathan Willington
James Grant
George John Sale
William Moray
Henry Yonge
Thomas Forster
Henry Walker
William Roycraft
James Conran
Lieutenants.—Edmund Safferey
William Brown
Hon. John Jones
James Read
John Burton
Frederick Willoe
Charles Johnson
Benjamin Adams
Thomas Lahiff
Edward Wrixon
John Brackenbury
H. E. Lynch
John D'Arcy
—— Johnson
William Gale
James Tomkinson
Michael Ryan

Lieutenants.—Joseph Budden
W. H. Robinson
F. W. Hutchinson
Cornets.—Thomas Kendall
Fran. Curtayne
Robert Willington
William Daniel
John Smith
J. M'Keale Anderson
Paymaster.—Robert Harman
Adjutant.—William Gale
Surgeon.—William King
Assistant-Surgeons.—John White
David Christie
Veterinary Surgeon.—Edward Coleman

1811

Colonel.—Oliver de Lancey
Lieut.-Colonels.—Hon. H. G. Grey
Evan Lloyd
William Carden
Majors.—Oswald Werge
Nathan Wilson
Captains.—David Supple
Philip Teesdale
Jonathan Willington
James Grant
G. J. Sale
William Moray
Henry Walker
William Roycraft
James Conran
William Brown
David M'Neale
Lieutenants.—Edmund Safferey
Hon. John Jones
John Burton
Frederick Willoe
Charles Johnson
Benjamin Adams
Thomas Lahiff
Edward Wrixon
John Brackenbury
Lieutenants.—H. E. Lynch
John D'Arcy
William Gale
—— Johnson
Michael Ryan
Joseph Budden
W. H. Robinson
Charles B. Sale
F. W. Hutchinson
Robert Coulthard
F. E. Cawne
John Smith
Thomas Kendall
Fran. Curtayne
Cornets.—Robert Willington
William Daniel
Henry Bond
J. M'Keale Anderson
Benjamin Astley
Isidore Blake
James Cockburn
Fra. Haworth
Paymaster.—Robert Harman
Adjutant.—William Gale
Quartermaster.—Thomas Carson
Surgeon.—William King
Assistant-Surgeons.—John White
David Christie
Veterinary Surgeon.—Edward Coleman

1812

Colonel.—Oliver de Lancey
Lieut.-Colonels.—Hon. H. G. Grey
Evan Lloyd
William Carden
Majors.—Oswald Werge
Nathan Wilson
Captains.—David Supple
Philip Teesdale
Jonathan Willington
James Grant
George John Sale
William Moray

Captains.—Henry Walker
William Roycraft
William Brown
Daniel M'Neale
John Burton

Lieutenants.—Hon. John Jones
Frederick Willoe
Charles Johnson
Benjamin Adams
Thomas Lahiff
Edward Wrixon
John Brackenbury
H. E. Lynch
John Darcy
William Gale
—— Johnson
Michael Ryan
Joseph Budden
W. H. Robinson
C. B. Sale
F. W. Hutchinson
Robert Coulthard
F. E. Cawne
John Smith
Thomas Kendall
Fran. Curtayne
James Cockburn
Robert Willington

Cornets.—William Daniel
J. M'K. Anderson
Benjamin Astley
Isidore Blake
Fran. Haworth
—— Carew
Samuel Orr
William MacFarlane
Samuel Enderby

Paymaster.—Robert Harman

Adjutant.—William Gale

Quartermaster.—Thomas Carson

Surgeon.—William King

Assistant-Surgeons.—John White
David Christie

Vet. Surgeon.—Edward Coleman

1813

Colonel.—Oliver de Lancey

Lieut.-Cols.—Evan Lloyd
William Carden
Hon. Lincoln Stanhope

Majors.—Oswald Werge
Nathan Wilson

Captains.—David Supple
Philip Teesdale
Jonathan Willington
James Grant
George Jno. Sale
William Moray
Henry Walker
William Roycraft
William Brown
Daniel M'Neale
Jno. Burton

Lieutenants.—Hon. John Jones
Frederick Willoe
Charles Johnson
Benjamin Adams
Thomas Lahiff
Edward Wrixon
John Brackenbury
Henry Edward Lynch
John D'Arcy
—— Johnson
Michael Ryan
Joseph Budden
W. H. Robinson
Charles Byrne Sale
F. W. Hutchinson
Robert Coulthard
F. E. Cawne
Fran. Curtayne
James Cockburn
Robert Willington
William Daniel
Henry Bond
Francis Haworth

Cornets.—J. M'Keale Anderson
Benjamin Astley
Isidore Blake

Cornets.—H. Carew
William MacFarlane
John Marks
Richard Willington
Paymaster.—Robert Harman
Adjutant.—John Marks
Quartermaster.—Thomas Carson
Surgeon.—William King
Assistant-Surgeon.—John Lorimer
Vet. Surgeon.—Edward Coleman

1814

Colonel.—Oliver de Lancey
Lieut.-Colonels.—Evan Lloyd
William Carden
Hon. L. Stanhope
Majors.—Oswald Werge
Nathan Wilson
Captains.—David Supple
Jonathan Willington
George John Sale
William Moray
Henry Walker
William Roycraft
Daniel M'Neale
James Burton
Hugh Percy Davidson
Hon. Leicester Stanhope
John Atkins
Lieutenants.—Hon. John Jones
Frederick Willoe
Charles Johnson
Benjamin Adams
Edward Wrixon
John Brackenbury
John D'Arcy
Michael Ryan
Joseph Budden
William H. Robinson
Charles Byrne Sale
F. W. Hutchinson
Robert Coulthard
F. E. Cawne
Francis Curtayne
Lieutenants.—James Cockburn
Robert Willington
William Daniel
Henry Bond
Francis Haworth
John Fraser
J. M'Keale Anderson
Benjamin Astley
Cornets.—Isidore Blake
H. Carew
W. MacFarlane
John Marks
Richard Willington
John Tomlinson
Thomas Hurring
William Gibson Peat
Oliver Delancey
Paymaster.—Robert Harman
Adjutant.—John Marks
Quartermaster.—Thomas Carson
Surgeon.—Alexander Young
Asst.-Surgeons.—John Lorimer
Eugene M'Swiney
Vet. Surgeon.—Edward Coleman

1815

Colonel.—Oliver de Lancey
Lieut.-Cols.—Evan Lloyd
William Carden
Hon. Lincoln Stanhope
Majors.—Oswald Werge
Nathan Wilson
Captains.—David Supple
Jonathan Willington
George John Sale
William Moray
Henry Walker
Daniel M'Neale
Hugh Percy Davidson
Hon. Leicester Stanhope
John Atkins
T. Perrouet Thompson
Joseph Smyth
Lieutenant.—Benjamin Adams

Lieutenants.—John Brackenbury
John D'Arcy
Michael Ryan
Joseph Budden
W. Henry Robinson
Charles Byrne Sale
F. W. Hutchinson
Robert Coulthard
Francis Curtayne
James Cockburn
Robert Willington
William Daniel
Henry Bond
Francis Haworth
Benjamin Astley
T. Ramsay Wharton
George Daun
C. G. A. Skinner
Isidore Blake
W. Hackett

Cornets.—H. Carew
William M'Farlane
Richard Willington
John Tomlinson
Thomas Hurring
W. Gibson Peat
Oliver de Lancey
William Potts
George Clarke
James Patch

Paymaster.—Robert Harman
Adjutant.—William Hackett
Quartermaster.—Thomas Carson
Surgeon.—Alexander Young
Asst.-Surgeons.—John Lorimer
Eugene M'Swiney
Vet. Surgeon.—Edward Coleman

1816

Colonel.—Oliver de Lancey
Lieut.-Cols.—Evan Lloyd
William Carden
Hon. Lincoln Stanhope
Major.—Oswald Werge
Major.—Nathan Wilson
Captains.—David Supple
Jonathan Willington
George John Sale
Daniel M'Neale
Hon. Leicester Stanhope
John Atkins
T. Perrouet Thompson
Benjamin Adams
Malcolm M'Neill

Lieutenants.—John Brackenbury
John D'Arcy
Joseph Budden
William H. Robinson
Charles Byrne Sale
F. W. Hutchinson
Robert Coulthard
Francis Curtayne
William Daniel
H. Bond
Francis Haworth
Isidore Blake
H. Carew
William M'Farlane
Samuel Ward Watson
William Hackett
John Tomlinson
Charles Greville

Cornets.—Richard Willington
Thomas Hurring
Oliver de Lancey
William Potts
George Clarke
James Patch
N. Raven
Thomas M'Kenzie
Peter Backhouse

Paymaster.—Robert Harman
Adjutant.—William Hackett
Quartermaster.—James Cockburn
Surgeon.—W. Wybrow
Asst.-Surgeons.—John Lorimer
Eugene M'Swiney
Vet. Surgeon.—Edward Coleman

Appendix A

1817

Colonel.—Oliver de Lancey
Lieut.-Cols.—Evan Lloyd
William Carden
Hon. Lincoln Stanhope
Majors.—Oswald Werge
Nathan Wilson
Captains.—David Supple
Jonathan Willington
George John Sale
Daniel M'Neale
John Atkins
Edward Byne
T. Perrouet Thompson
Benjamin Adams
Malcolm M'Neill
Lieutenants.—John Brackenbury
John D'Arcy
Joseph Budden
W. H. Robinson
Charles Byrne Sale
F. W. Hutchinson
Robert Coulthard
Francis Curtayne
William Daniel
Henry Bond
Francis Haworth
Isidore Blake
H. Carew
W. M'Farlane
Samuel Ward Watson
Richard Willington
Ambrose de L'Etang
John Tomlinson
Henry Court Amiel
Charles Greville
T. L. Stuart Menteath
Cornets.—Thomas Hurring
Oliver de Lancey
William Potts
George Clarke
T. Ellman
J. Patch
N. Raven
Cornets.—P. Backhouse
Thomas Carey
Thomas Nicholson
Paymaster.—Robert Harman
Adjutant.—Thomas Carey
Quartermaster.—James Cockburn
Surgeon.—William Wybrow
Asst.-Surgeons.—John Lorimer
Thomas Price
Vet. Surgeon.—Edmund Price

1818

Colonel.—Oliver de Lancey
Lieut.-Cols.—Evan Lloyd
William Carden
Hon. Lincoln Stanhope
Majors.—Oswald Werge
Nathan Wilson
Captains.—David Supple
Jonathan Willington
George John Sale
Daniel M'Neale
John Atkins
Edward Byne
T. Perrouet Thompson
Benjamin Adams
Malcolm M'Neill
Charles Wayth
Lieutenants.—John Brackenbury
John D'Arcy
Joseph Budden
W. Henry Robinson
Charles Byrne Sale
F. W. Hutchinson
Robert Coulthard
Francis Curtayne
William Daniel
Henry Bond
Isidore Blake
H. Carew
William M'Farlane
Samuel Ward Watson
Richard Willington
Ambrose de L'Etang

Lieutenants.—John Tomlinson
Henry Court Amiel
T. L. Stuart Menteath
Thomas Hurring
Oliver de Lancey
Cornets.—William Potts
George Clarke
T. Ellman
James Patch
N. Raven
Peter Backhouse
Thomas Nicholson
James Byrne Smith
J. B. Nixon
Paymaster.—Robert Harman
Adjutant.—James Byrne Smith
Quartermaster.—James Cockburn
Surgeon.—William Wybrow
Asst.-Surgeons.—John Lorimer
Thomas Price
Vet. Surgeon.—Edmund Price

1819

Colonel.—Oliver de Lancey
Lieut.-Cols.—Evan Lloyd
Hon. L. Stanhope
Oswald Werge
Majors.—Nathan Wilson
Jonathan Willington
Captains.—George John Sale
Daniel M'Neale
John Atkins
Edward Byne
T. Perrouet Thompson
Benjamin Adams
Malcolm M'Neill
Charles Wayth
John Brackenbury
Lieutenants.—John D'Arcy
Joseph Budden
W. Henry Robinson
F. W. Hutchinson
Francis Curtayne
William Daniel
Lieutenants.—Henry Bond
Isidore Blake
H. Carew
William M'Farlane
Samuel Ward Watson
Richard Willington
Ambrose de L'Etang
John Tomlinson
Henry Court Amiel
T. L. Stuart Menteath
Thomas Hurring
Oliver de Lancey
W. T. H. Fisk
Cornets.—William Potts
George Clarke
T. Ellman
N. Raven
Peter Backhouse
Thomas Nicholson
John Byrne Smith
J. B. Nixon
William Marriott
Paymaster.—Robert Harman
Adjutant.—J. R. Smith
Quartermaster.—James Cockburn
Surgeon.—W. Wybrow
Acting-Surgeons.—John Lorimer
Thomas Price
Vet. Surgeon.—Edmund Price

1820

Colonel.—Oliver de Lancey
Lieut.-Colonels.—Evan Lloyd
Hon. L. Stanhope
Oswald Werge
Majors.—Nathan Wilson
Jonathan Willington
Captains.—George John Sale
Dan. M'Neale
John Atkins
Edward Byne
Thomas P. Thompson
Benjamin Adams
Malcolm M'Neill

Captains.—Charles Wayth
John Brackenbury
Lieutenants.—John D'Arcy
Joseph Budden
W. H. Robinson
Charles Byrne Sale
F. W. Hutchinson
Francis Curtayne
William Daniel
Henry Bond
Isidore Blake
H. Carew
Wm. M'Farlane
Richard Willington
Ambrose de L'Etang
H. Court Amiel
T. L. Stuart Menteath
Thomas Hurring
Oliver de Lancey
William T. H. Fisk
George F. Clarke
George G. Shaw
Cornets.—William Potts
N. Raven
Peter Backhouse
Thomas Nicholson
James Byrne Smith
William Marriott
Charles St. John Fancourt
Frederick Loftus
Paymaster.—Robert Harman
Adjutant.—James Byrne Smith
Quartermaster.—James Cockburn
Surgeon.—William Wybrow
Assistant-Surgeons.—John Lorimer
Thomas Price
Veterinary Surgeon.—Edmund Price

1821

Colonel.—Oliver de Lancey
Lieut.-Colonels.—Evan Lloyd
Hon. L. Stanhope
Nathan Wilson
Major.—Jonathan Willington
Major.—George John Sale
Captains.—Daniel M'Neale
John Atkins
Edward Byne
Thomas P. Thompson
Benjamin Adams
Malcolm M'Neill
Charles Wayth
John Brackenbury
William H. Robinson
Lieutenants.—John D'Arcy
Joseph Budden
Charles Byrne Sale
Francis Curtayne
William Daniel
Henry Bond
Isidore Blake
H. Carew
William M'Farlane
Richard Willington
Ambrose de L'Etang
Henry Court Amiel
T. L. S. Menteath
Thomas Hurring
W. T. Hawley Fisk
George F. Clarke
George G. Shaw
W. H. B. Lindsay
N. Raven
Cornets.—W. Potts
Peter Backhouse
Thomas Nicholson
Robert Lewis
Charles St. John Fancourt
Frederick Loftus
Arch. Edmund Bromwich
Hon. Nat. Hen. Chas. Massey
Paymaster.—Robert Harman
Quartermaster.—James Cockburn
Surgeon.—William Wybrow
Assistant-Surgeons.—John Lorimer
Samuel Holmes
Veterinary Surgeon.—Edmund Price

1822

Colonel.—Oliver de Lancey
Lieut.-Colonels.—Evan Lloyd
Hon. L. Stanhope
Nathan Wilson
Majors.—Jonathan Willington
Norcliffe Norcliffe
Captains.—Daniel M'Neale
John Atkins
Edward Byne
Thomas P. Thompson
Benjamin Adams
Malcolm M'Neill
Charles Wayth
John Brackenbury
William H. Robinson
Lieutenants.—John D'Arcy
Joseph Budden
Charles Byrne Sale
Francis Curtayne
William Daniel
Henry Bond
Isidore Blake
H. Carew
William M'Farlane
Richard Willington
Henry Court Amiel
T. L. S. Menteath
Thomas Hurring
W. T. Hawley Fisk
George G. Shaw
N. Raven
W. Potts
Cornets.—Peter Backhouse
Thomas Nicholson
Robert Lewis
C. St. John Fancourt
Frederick Loftus
Arch. E. Bromwich
William Penn
Hon. Nat. Hen. Chas. Massey
Paymaster.—Robert Harman
Adjutant.—W. T. Hawley Fisk
Quartermaster.—James Cockburn
Surgeon.—William Wybrow
Assistant-Surgeons.—John Lorimer
Sam. Holmes
Veterinary Surgeon.—Edmund Price

1823

Colonel.—Lord R. E. H. Somerset, K.C.B.
Lieut.-Colonels.—Evan Lloyd
Hon. L. Stanhope
Majors.—Jonathan Willington
Norcliffe Norcliffe
Captains.—Daniel M'Neale
John Atkins
Edward Byne
Thomas P. Thompson
Benjamin Adams
Malcolm M'Neill
John Brackenbury
William H. Robinson
W. T. Cockburn
Lieutenants.—John D'Arcy
Joseph Budden
Charles Byrne Sale
Francis Curtayne
Henry Bond
Isidore Blake
H. Carew
William M'Farlane
Rich. Willington
Henry Court Amiel
Thomas Hurring
W. T. Hawley Fisk
George G. Shaw
N. Raven
William Potts
William Graham
Cornets.—Peter Backhouse
Thomas Nicholson
Robert Lewis
Frederick Loftus
Arch. Edmund Bromwich
William Penn

Cornets.—Hon. Nat. H. C. Massey
Lewis Shedden
Paymaster.—Robert Harman
Adjutant.—W. T. Hawley Fisk
Quartermaster.—James Cockburn
Surgeon.—William Wybrow
Asst.-Surgs.—John Lorimer, M.D.
Sam. Holmes, M.D.
Veterinary Surgeon.—Edmund Price

1824

Colonel.—Lord R. E. H. Somerset, K.C.B.
Lieut.-Colonels.—Evan Lloyd
Hon. L. Stanhope
Majors.—J. Willington
George Luard
Captains.—Daniel M'Neale
Thomas P. Thompson
Benjamin Adams
Malcolm M'Neill
John Brackenbury
John Scott
Lieutenants.—John D'Arcy
Joseph Budden
Harry Bond
W. T. Hawley Fisk
George F. Clarke
George Robbins
William Dungan
Thomas Nicholson
Cornets.—Robert Lewis
Frederick Loftus
William Penn
Hon. N. H. C. Massey
Samuel Pole
R. J. Elton
Paymaster.—Robert Harman
Adjutant.—W. T. H. Fisk
Quartermaster.—James Cockburn
Surgeon.—William Wybrow
Assistant-Surgeon.—John Lorimer
Veterinary Surgeon.—Edmund Price
Agents.—Hopkinson & Sons

1825

Colonel.—Lord R. E. H. Somerset, K.C.B.
Lieut.-Colonels.—Evan Lloyd
Hon. L. Stanhope
Majors.—J. Willington
George Luard
Captains.—T. P. Thompson
Benjamin Adams
J. Brackenbury
John Scott
William Locke
Frederick Johnston
Lieutenants.—John D'Arcy
Joseph Budden
W. T. Hawley Fisk
George F. Clarke
George Robbins
William Dungan
George T. Greenland
M. C. D. St. Quintin
Cornets.—Frederick Loftus
Hon. N. H. C. Massey
Samuel Pole
R. J. Elton
John Barron
Hon. R. F. Greville
Paymaster.—Robert Harman
Adjutant.—W. T. H. Fisk
Quartermaster.—T. Nicholson
Surgeon.—William Wybrow
Assistant-Surgeon.—John Lorimer
Veterinary Surgeon.—Henry Smith

1826

Colonel.—Lord R. E. H. Somerset, K.C.B.
Lieut.-Colonels.—Evan Lloyd
Hon. L. Stanhope
Majors.—George Luard
Lord Bingham
Captains.—Benjamin Adams
John Scott
Frederick Johnston

Captains.—W. N. Burrows
George F. Clarke
Alan Chambre
Lieutenants.—W. T. H. Fisk
George Robbins
William Dungan
G. T. Greenland
M. C. D. St. Quintin
Frederick Loftus
Hon. Nat. Hen. Chas. Massey
Samuel Pole
Cornets.—R. J. Elton
John Barron
Hon. R. F. Greville
Charles Forbes
Henry Witham
S. J. W. F. Welch
Paymaster.—Robert Harman
Adjutant.—W. T. H. Fisk
Quartermaster.—T. Nicholson
Surgeon.—William Wybrow
Assistant-Surgeon.—Sam. Holmes
Veterinary Surgeon.—Henry Smith

1827

Colonel.—Lord R. E. H. Somerset, K.C.B.
Lt.-Cols.—Evan Lloyd
George, Lord Bingham
Majors.—Anthony Bacon
John Scott
Captains.—William N. Burrowes
George F. Clarke
George Robbins
George T. Greenland
M. C. D. St. Quintin
George M. Keane
Lieutenants.—Robert James Elton
John Barron
Charles Forbes
Henry Witham
S. J. W. F. Welch
Cornet.—Nat. B. F. Shawe
Cornets.—Samuel W. Need
W. C. Douglas
William Murray Percy
William Henry Tonge
Lionel Ames
Paymaster.—W. T. Hawley Fisk
Adjutant.—John Barron
Quartermaster.—T. Nicholson
Surgeon.—William Wybrow
Assistant-Surgeon.—H. G. Parken, M.D.
Vet. Surgeon.—John Wilkinson

1828

Colonel.—Lord R. E. H. Somerset, K.C.B.
Lt.-Cols.—Evan Lloyd
George, Lord Bingham
Majors.—John Scott
William N. Burrowes
Captains.—George F. Clarke
George Robbins
M. C. D. St. Quintin
John Lawrenson
Robert James Elton
Lieutenants.—John Barron
Charles Forbes
Henry Witham
Nat. B. F. Shawe
W. C. Douglas
Samuel Need
William M. Percy
Cornets.—William H. T. Tonge
Lionel Ames
A. H. Mitchelson
Denis Hanson
William Wentworth
William L. Shedden
Paymaster.—W. T. Hawley Fisk
Adjutant.—Denis Hanson
Quartermaster.—T. Nicholson
Surgeon.—William Wybrow
Asst.-Surgeon.—H. G. Parken, M.D.
Vet. Surgeon.—John Wilkinson

Appendix A

1829

Colonel.—Lord R. E. H. Somerset, K.C.B.
Lt.-Cols.—Evan Lloyd
George, Lord Bingham
Majors.—John Scott
W. N. Burrowes
Captains.—George F. Clarke
George Robbins
M. C. D. St. Quintin
George M. Keane
John Lawrenson
Robert James Elton
Lieutenants.—John Barron
Charles Forbes
Harry Witham
N. B. F. Shawe
William C. Douglas
Samuel W. Need
William M. Percy
Cornets.—William H. Tonge
Lionel Ames
A. H. Michelson
Denis Hanson
William Wentworth
W. L. Shedden
Paymaster.—G. Chandler
Adjutant.—Denis Hanson
Quartermaster.—T. Nicholson
Surgeon.—James G. Elkington
Assistant-Surgeon.—H. G. Parken
Vet. Surgeon.—John Wilkinson
Agent.—Mr. Hopkinson

1830

Colonel.—Sir J. Elley, K.C.B.
Lt.-Cols.—Evan Lloyd
George, Lord Bingham
Majors.—John Scott
W. N. Burrowes
Captains.—George F. Clarke
George Robbins
M. C. D. St. Quintin
George M. Keane
John Lawrenson
Robert K. Trotter
Lieutenants.—John Barron
Charles Forbes
N. B. F. Shawe
Samuel W. Need
William C. Douglas
William M. Percy
William H. Tonge
Cornets.—Lionel Ames
Denis Hanson
W. L. Shedden
H. F. Walker
Walter Williams
Philip J. West
Paymaster.—G. Chandler
Adjutant.—Denis Hanson
Quartermaster.—Thos. Nicholson
Surgeon.—James G. Elkington
Asst.-Surgeon.—H. G. Parken
Vet. Surgeon.—John Wilkinson
Agent.—Mr. Hopkinson

1831

Colonel.—Sir J. Elley, K.C.B.
Lt.-Cols.—Sir Evan Lloyd
George, Lord Bingham
Major.—W. N. Burrowes
Captains.—George F. Clarke
George Robbins
M. C. D. St. Quintin.
George M. Keane
John Lawrenson
Robert R. Trotter
Lieutenants.—John Barron
Charles Forbes
N. B. F. Shawe
Samuel W. Need
W. C. Douglas
W. M. Percy
W. H. Tonge
Cornets.—Lionel Ames
Denis Hanson
W. L. Shedden

Cornets.—H. F. Walker
Walter Williams
Philip J. West
Paymaster.—G. Chandler
Adjutant.—Denis Hanson
Surgeon.—J. G. Elkington
Asst.-Surgeon.—H. G. Parken
Vet. Surgeon.—John Wilkinson
Quartermaster.—Thos. Nicholson

1832

Colonel.—Sir J. Elley, K.C.B.
Lt.-Cols.—Sir Evan Lloyd
George, Lord Bingham
Major.—W. N. Burrowes
Captains.—George F. Clarke
George Robbins
M. C. D. St. Quintin
George M. Keane
John Lawrenson
Robert K. Trotter
Lieutenants.—Charles Forbes
N. B. F. Shawe
Samuel W. Need
W. C. Douglas
W. M. Percy
W. H. Tonge
Lionel Ames
Cornets.—Denis Hanson
W. L. Shedden
W. Williams
P. J. West
F. J. Parry
W. H. Fielden
Paymaster.—G. Chandler
Adjutant.—Denis Hanson
Surgeon.—J. G. Elkington
Asst.-Surgeon.—H. G. Parken
Vet. Surgeon.—John Wilkinson
Quartermaster.—William Hall

1833

Colonel.—Sir J. Elley, K.C.B.
Lt.-Col.—Sir Evan Lloyd
Lt.-Col.—George, Lord Bingham
Major.—Henry Pratt
Captains.—George Robbins
M. C. D. St. Quintin
George M. Keane
John Lawrenson
Robert K. Trotter
Charles Forbes
Lieutenants.—N. B. F. Shawe
Samuel W. Need
W. C. Douglas
Lionel Ames
Denis Hanson
W. L. Shedden
Walter Williams
Cornets.—Philip West
F. J. Parry
W. H. Fielden
Edward Croker
R. W. Macdonald
R. A. F. Kingscote
Paymaster.—G. Chandler
Adjutant.—Denis Hanson
Quartermaster.—William Hall
Surgeon.—J. G. Elkington
Asst.-Surgeon.—H. G. Parken
Vet. Surgeon.—John Wilkinson

1834

Colonel.—Sir J. Elley, K.C.B.
Lt.-Cols.—Sir Evan Lloyd
George, Lord Bingham
Major.—Henry Pratt
Captains.—M. C. D. St. Quintin
George M. Keane
John Lawrenson
R. K. Trotter
Charles Forbes
N. B. F. Shawe
Lieutenants.—Samuel W. Need
W. C. Douglas
Lionel Ames
Denis Hanson
W. L. Shedden

Lieutenants.—W. Williams
P. J. West
Cornets.—F. J. Parry
W. H. Fielden
Edward Croker
R. W. Macdonald
R. A. F. Kingscote
John Mordaunt
Paymaster.—G. Chandler
Adjutant.—Denis Hanson
Quartermaster.—William Hall
Surgeon.—J. Elkington
Asst.-Surgeon.—H. G. Parken
Vet.-Surgeon.—John Wilkinson

1835

Colonel.—Sir J. Elley, K.C.B
Lt.-Cols.—Sir Evan Lloyd
George, Lord Bingham
Major.—Henry Pratt
Captains.—M. C. D. St. Quintin
George M. Keane
John Lawrenson
K. R. Trotter
Charles Forbes
N. B. F. Shawe
Lieutenants.—Samuel W. Need
W. C. Douglas
Lionel Ames
Denis Hanson
W. L. Shedden
W. Williams
P. J. West
Cornets.—F. J. Parry
W. H. Fielden
Edward Croker
R. W. M'Donald
R. A. F. Kingscote
John Mordaunt
Paymaster.—G. Chandler
Adjutant.—Denis Hanson
Quartermaster.—William Hall
Surgeon.—J. G. Elkington
Asst.-Surgeon.—H. G. Parken
Vet. Surgeon.—John Wilkinson

1836

Colonel.—Sir J. Elley, K.C.B.
Lt.-Cols.—Sir Evan Lloyd
George, Lord Bingham
Major.—Henry Pratt
Captains.—M. C. D. St. Quintin
G. M. Keane
John Lawrenson
R. K. Trotter
N. B. F. Shawe
W. C. Douglas
Lieutenants.—Lionel Ames
Denis Hanson
W. L. Shedden
W. Williams
W. H. Fielden
Edward Croker
R. W. Macdonald
Cornets.—R. A. F. Kingscote
John Mordaunt
Wallace Barrow
J. R. Palmer
J. B. Broadley
Robert Reynard
Paymaster.—George Chandler
Adjutant.—Denis Hanson
Quartermaster.—William Hall
Surgeon.—J. G. Elkington
Asst.-Surgeon.—H. G. Parken
Vet. Surgeon.—John Wilkinson

1837

Colonel.—Sir J. Elley, K.C.B.
Lt.-Cols.—Sir Evan Lloyd
George, Lord Bingham
Major.—Henry Pratt
Captains.—M. C. D. St. Quintin
George M. Keane
John Lawrenson
W. C. Douglas
Lionel Ames
W. L. Shedden

Lieutenants.—Denis Hanson
W. Williams
W. H. Fielden
Edward Croker
R. A. F. Kingscote
John Mordaunt
Wallace Barrow
Cornets.—J. R. Palmer
J. R. Broadley
Robert Reynard
John D. Brett
William M. Mitchell
A. S. Willett
Paymaster.—G. Chandler
Adjutant.—Denis Hanson
Quartermaster.—William Hall
Surgeon.—J. G. Elkington
Asst.-Surgeon.—J. B. Gibson, M.D.
Vet. Surgeon.—John Wilkinson

1838

Colonel.—Sir J. Elley, K.C.B.
Lieut.-Colonel.—Henry Pratt
Major.—M. C. D. St. Quintin
Captains.—John Lawrenson
W. C. Douglas
Lionel Ames
W. L. Shedden
W. Williams
W. H. Fielden
Lieutenants.—Edward Croker
R. A. F. Kingscote
W. Barrow
J. R. Palmer
J. B. Broadley
R. A. Houblon
Francis Burdett
Cornets.—Robert Reynard
J. D. Brett
W. M. Mitchell
A. S. Willett
Hon. G. O'Callaghan
Andrew Wauchope
Paymaster.—Captain G. Chandler
Adjutant.—Wallace Barrow
Quartermaster.—William Hall
Surgeon.—J. G. Elkington
Asst.-Surgeon.—J. B. Gibson, M.D.
Vet. Surgeon.—John Wilkinson

1839

Colonel.—Sir J. Elley, K.C.B.
Lieut.-Colonel.—Henry Pratt
Major.—M. C. D. St. Quintin
Captains.—J. Lawrenson
W. C. Douglas
Lionel Ames
W. L. Shedden
W. Williams
W. H. Fielden
Lieutenants.—Edward Croker
R. A. F. Kingscote
W. Barrow
J. R. Palmer
J. B. Broadley
Richard A. Houblon
Francis Burdett
Cornets.—Robert Reynard
J. D. Brett
W. M. Mitchell
A. S. Willett
Hon. G. O'Callaghan
Andrew Wauchope
Paymaster.—G. Chandler
Adjutant.—Wallace Barrow
Quartermaster.—William Hall
Surgeon.—J. G. Elkington
Asst.-Surgeon.—J. B. Gibson, M.D.
Vet. Surgeon.—John Wilkinson
Agents.—Hopkinson & Sons

1840

Colonel.—Sir A. B. Clifton, K.C.B.
Lt.-Col.—M. C. D. St. Quintin
Major.—John Lawrenson
Captains.—William C. Douglas
Lionel Ames
Walter Williams

Captains.—Edmund Croker
R. A. F. Kingscote
Wallace Barrow
Lieutenants.—J. R. Palmer
J. B. Broadley
Francis Burdett
J. D. Brett
Archibald, Earl of Cassilis
W. M. Mitchell
Aug. Saltern Willett
Cornets.—Thomas Lindsay
Edward C. Scobell
H. R. Boucherett
Abraham Hamilton
William O. Hammond
H. Roxby Benson
Paymaster.—G. Chandler
Adjutant.—Thomas Lindsay
Quartermaster.—William Hall
Surgeon.—James G. Elkington
Assistant-Surgeon.—J. B. Gibson, M.D.
Veterinary Surgeon.—J. Wilkinson

1841

Colonel.—Sir A. B. Clifton, K.C.B.
Lieutenant-Colonel.—M. C. D. St. Quintin
Major.—John Lawrenson
Captains.—William C. Douglas
Walter Williams
Edward Croker
R. A. F. Kingscote
Wallace Barrow
J. R. Palmer
Lieutenants.—J. B. Broadley
Francis Burdett
J. D. Brett
Archibald, Earl of Cassilis
A. S. Willett
Hon. H. S. Blackwood
Thomas Lindsay
Lieutenant.—E. C. Scobell
Cornets.—H. R. Boucherett
Abraham Hamilton
William O. Hammond
H. R. Benson
Charles W. Miles
Paymaster.—G. Chandler
Adjutant.—Thomas Lindsay
Quartermaster.—William Hall
Surgeon.—James G. Elkington
Assistant-Surgeon.—J. B. Gibson, M.D.
Veterinary Surgeon.—J. Wilkinson

1842

Colonel.—Sir A. B. Clifton, K.C.B.
Lt.-Col.—M. C. D. St. Quintin
Major.—John Lawrenson
Captains.—W. C. Douglas
Walter Williams
R. A. F. Kingscote
J. R. Palmer
J. B. Broadley
Francis Burdett
Lieutenants.—J. D. Brett
Archibald, Earl of Cassilis
A. S. Willett
Hon. H. S. Blackwood
Thomas Lindsay
Edward C. Scobell
H. R. Boucherett
Abraham Hamilton
Cornets.—W. O. Hammond
H. R. Benson
C. W. Miles
Wm. A., Lord Inverury
H. C. Taylor
Paymaster.—George Chandler
Adjutant.—Thomas Lindsay
Quartermaster.—William Hall
Surgeon.—Edward Pilkington
Assistant-Surgeon.—Alex. Leslie
Veterinary Surgeon.—J. Wilkinson

1843

Colonel.—H.R.H. Prince George of Cambridge
Lt.-Col.—M. C. D. St. Quintin
Major.—John Lawrenson
Captains.—J. R. Palmer
John B. Broadley
Francis Burdett
J. D. Brett
A. S. Willett
Hon. H. S. Blackwood
Lieutenants.—Thomas Lindsay
E. C. Scobell
H. R. Boucherett
Abraham Hamilton
H. R. Benson
Charles W. Miles
Wm. A., Lord Inverury
Cornets.—H. C. Taylor
Alfred Crawshay
Thomas Lyon
Samuel Le H. Hodson
N. M. Innes
Paymaster.—George Chandler
Adjutant.—Thomas Lindsay
Quartermaster.—William Hall
Surgeon.—Edward Pilkington
Assistant-Surgeon.—G. Anderson
Vet. Surgeon.—John Wilkinson

1844

Colonel.—H.R.H. Prince George of Cambridge
Lt.-Col.—M. C. D. St. Quintin
Major.—John Lawrenson
Captains.—J. R. Palmer
J. B. Broadley
Francis Burdett
J. D. Brett
A. S. Willett
E. C. Scobell
Lieutenants.—Thomas Lindsay
H. R. Boucherett
Abraham Hamilton
H. R. Benson
C. W. Miles
H. C. Taylor
Alfred Crawshay
Thomas Lyon
Cornets.—Samuel Le H. Hobson
N. M. Innes
J. F. Blathwayt
E. C. A. Haworth
R. D. Hay Lane
Paymaster.—George Chandler
Adjutant.—H. T. Lindsay
Quartermaster.—William Hall
Surgeon.—Edward Pilkington
Assistant-Surgeon.—G. Anderson
Vet. Surgeon.—John Wilkinson

1845

Colonel.—H.R.H. Prince George of Cambridge
Lt.-Col.—M. C. D. St. Quintin
Major.—John Lawrenson
Captains.—J. R. Palmer
Francis Burdett
John D. Brett
A. S. Willett
E. C. Scobell
H. R. Boucherett
Lieutenants.—Abraham Hamilton
H. R. Benson
Charles W. Miles
Alfred Crawshay
Thomas Lyon
Norman M. Innes
J. E. Fleeming
Cornets.—E. C. A. Haworth
J. F. Blathwayt
R. D. Hay Lane
John Stephenson
Henry W. Lindow
William I. Anderton
Paymaster.—George Chandler
Adjutant.—John Stephenson

Quartermaster.—William Hale
Surgeon.—John Brown Gibson, M.D.
Assistant-Surgeon.—G. Anderson
Vet. Surgeon.—John Wilkinson

1846

Colonel.—H.R.H. Prince George of Cambridge
Lt.-Col.—M. C. D. St. Quintin
Major.—Francis Burdett
Captains.—John D. Brett
A. S. Willett
E. C. Scobell
H. R. Boucherett
Abraham Hamilton
H. R. Benson
Lieutenants.—Charles W. Miles
Alfred Crawshay
Thomas Lyon
J. E. Fleeming
E. C. A. Haworth
R. D. Hay Lane
John Stephenson
W. I. Anderton
Cornets.—J. C. W. Russell
E. R. Dodwell
P. J. W. Miles
W. W. Codrington
William H. K. Erskine
Paymaster.—George Chandler
Adjutant.—John Stephenson
Quartermaster.—Wm. Hall
Surgeon.—J. B. Gibson, M.D.
Asst.-Surgeon.—H. Kendall, M.D.
Veterinary Surgeon.—W. C. Lord

1847

Colonel.—H.R.H. Prince George of Cambridge
Lt.-Col.—M. C. D. St. Quintin
Major.—Francis Burdett
Captains.—John D. Brett
A. S. Willett
E. C. Scobell
H. R. Boucherett
Abraham Hamilton
H. R. Benson
Lieutenants.—Charles W. Miles
Alfred Crawshay
Thomas Lyon
J. E. Fleeming
E. C. A. Haworth
R. D. Hay Lane
John Stephenson
William I. Anderton
Cornets.—J. C. W. Russell
E. R. Dodwell
Philip J. W. Miles
W. W. Codrington
William H. K. Erskine
Paymaster.—George Chandler
Adjutant.—John Stephenson
Quartermaster.—William Hall
Surgeon.—J. B. Gibson, M.D.
Asst.-Surgeon.—H. Kendall, M.D.
Veterinary Surgeon.—W. C. Lord

1848

Colonel.—H.R.H. Prince George of Cambridge
Lt.-Col.—M. C. D. St. Quintin
Major.—Francis Burdett
Captains.—John D. Brett
A. S. Willett
Abraham Hamilton
H. R. Benson
C. W. Miles
Thomas Lyon
Lieutenants.—J. E. Fleeming
E. C. A. Haworth
R. D. Hay Lane
W. I. Anderton
William Morris
J. C. W. Russell
Philip J. W. Miles
W. W. Codrington
Cornet.—W. H. K. Erskine

Cornets.—H. St. George, R.M.
Alexander Campbell
William F. Webb
Robert White
Paymaster.—John Stephenson
Adjutant.—J. E. Fleeming
Quartermaster.—William Hall
Surgeon.—J. B. Gibson, M.D.
Asst.-Surgeon.—H. Kendall, M.D.
Veterinary Surgeon.—William C. Lord

1849

Colonel.—H.R.H. Prince George of Cambridge
Lieutenant-Colonel.—M. C. D. St. Quintin
Major.—Francis Burdett
Captains.—John Dary Brett
A. S. Willett
Abraham Hamilton
H. R. Benson
J. E. Fleeming
E. C. A. Haworth
Lieutenants.—R. D. Hay Lane
W. I. Anderton
William Morris
J. C. W. Russell
W. H. R. Erskine
Howard St. George
W. F. Richards
William F. Webb
Robert White
Cornets.—J. P. Winter
Thomas Taylor, R.M.
J. H. Reed
A. F. C. Webb
Paymaster.—John Stephenson
Adjutant.—Howard St. George
Quartermaster.—William Hall
Surgeon.—J. B. Gibson, M.D.
Asst.-Surgeon.—H. Kendall, M.D.
Veterinary Surgeon.—William C. Lord

1850

Colonel.—H.R.H. Prince George of Cambridge
Lt.-Colonel.—M. C. D. St. Quintin
Major.—Francis Burdett
Captains.—John D. Brett
A. S. Willett
Abraham Hamilton
H. R. Benson
J. E. Fleeming
E. C. A. Haworth
Lieutenants.—R. D. Hay Lane
William Morris
J. C. W. Russell
W. H. K. Erskine
Howard St. George
W. F. Richards
Robert White
John Pratt Winter
Joseph H. Reed
Cornets.—Thomas Taylor, R.M.
A. F. C. Webb
Godfrey C. Morgan
A. Learmonth
Paymaster.—John Stephenson
Adjutant.—Howard St. George
Quartermaster.—William Hall
Surgeon.—J. B. Gibson, M.D.
Asst.-Surg.—Henry Kendall, M.D.
Vet. Surgeon.—William C. Lord

1851

Colonel.—H.R.H. Duke of Cambridge
Lt.-Colonel.—M. C. D. St. Quintin
Major.—Francis Burdett
Captains.—John D. Brett
A. S. Willett
Abraham Hamilton
H. R. Benson
E. C. A. Haworth
R. D. Hay Lane
Lieutenants.—William Morris
W. H. K. Erskine

Lieutenants.—Howard St. George
W. F. Richards
Robert White
John Pratt Winter
A. F. C. Webb
G. C. Morgan
A. Learmonth
Cornets.—Thomas Taylor, R.M.
John Henry Thomson
Sir W. Gordon, Bart.
Lewis Edward Knight
Paymaster.—John Stephenson
Adjutant.—Howard St. George
Quartermaster.—William Hall
Surgeon.—J. B. Gibson, M.D.
Asst.-Surg.—Henry Kendall, M.D.
Vet. Surgeon.—William C. Lord

1852

Colonel.—H.R.H. Duke of Cambridge, K.G.
Lieut.-Colonel.—John Lawrenson
Major.—John D. Brett
Captains.—A. S. Willett
H. R. Benson
E. C. A. Haworth
William Morris
W. H. K. Erskine
W. Fred. Richards
Lieutenants.—Robert White
John Pratt Winter
A. F. C. Webb
G. C. Morgan
A. Learmonth
John H. Thompson
Sir W. Gordon, Bart.
Lewis E. Knight.
W. F. Tollemache
Cornets.—Thomas Taylor, R.M.
John Thomas Cator
George Ross
J. W. Cradock-Hartopp
Paymaster.—J. Stephenson
Adjutant.—A. Learmonth
Quartermaster.—W. Hall
Surgeon.—J. B. Gibson, M.D.
Asst.-Surgeon.—H. Kendall, M.D.
Vet. Surgeon.—W. C. Lord

1853

Colonel.—T. W. Taylor, C.B.
Lieut.-Colonel.—John Lawrenson
Major.—A. S. Willett
Captains.—H. R. Benson
Wm. Morris
Wm. H. K. Erskine
John Pratt Winter
A. F. C. Webb
Lieutenants.—G. C. Morgan
A. Learmonth
J. H. Thompson
Sir W. Gordon, Bart.
Lewis E. Knight
Wm. F. Tollemache
Cornets.—Thos. Taylor, R.M.
J. W. Cradock-Hartopp
John Chadwick
Philip Musgrave
W. J. Pearson Watson
Sir G. H. Leith, Bart.
G. O. Wombwell
Paymaster.—John Stephenson
Adjutant.—John Chadwick
Quartermaster.—John Yates
Surgeon.—J. B. Gibson, M.D.
Asst.-Surgeon.—H. Kendall, M.D.
Vet. Surgeon.—S. Price Constant

1854

Colonel.—T. W. Taylor, C.B.
Lieut.-Colonel.—J. Lawrenson
Major.—A. S. Willett
Captains.—H. R. Benson
Wm. Morris
Robert White
J. Pratt Winter
A. F. C. Webb
Godfrey C. Morgan

Lieutenants.—A. Learmonth
J. H. Thompson
Sir W. Gordon, Bart.
Lewis E. Knight
J. W. Cradock-Hartopp
Philip Musgrave
Cornets.—Thos. Taylor, R.M.
J. Chadwick
W. J. Pearson Watson
Sir G. H. Leith, Bart.
G. O. Wombwell
Archibald Cleveland
A. F. S. Jerningham
Paymaster.—J. Stephenson
Adjutant.—J. Chadwick
Quartermaster.—John Yates
Surgeon.—J. B. Gibson, M.D.
Asst.-Surgeon.—H. Kendall, M.D.
Vet. Surgeon.—S. P. Constant

1855

Colonel.—Sir. J. M. Wallace, K.H.
Lieut.-Colonel.—J. Lawrenson
Major.—Henry R. Benson
Captains.—Wm. Morris
Robert White
Godfrey C. Morgan
Alex. Learmonth
Sir Wm. Gordon, Bart.
Lewis Edward Knight
J. W. C. Hartopp
John Macartney
Lieutenants.—W. J. P. Watson
Thos. Taylor, R.M.
John Chadwick
Sir G. H. Leith, Bart.
G. O. Wombwell
Drury Curzon Lowe
Arthur Burnand
Henry H. Barber
Henry Baring
Cornets.—G. H. L. Boynton
Wm. D. Nath. Lowe
Cornets.—Wm. Digby Seymour
John Gibsone
Paymaster.—John Stephenson
Adjutant.—John Chadwick
Quartermaster.—C. J. Ffennell
Surgeon.—H. H. Massey, M.D.
Asst.-Surgeon.—St. John Stanley
Vet. Surgeon.—S. P. Constant

1856

Colonel.—Sir J. M. Wallace, K.H.
Lieut.-Colonel.—John Lawrenson
Major.—Henry R. Benson
Captains.—Wm. Morris, C.B. (Major)
Robert White
Alex. Learmonth
Sir W. Gordon, Bart.
Lewis Edward Knight
John Macartney
W. J. P. Watson
Sir G. H. Leith, Bart.
Lieutenants.—Thos. Taylor, R.M.
John Chadwick
Drury Curzon Lowe
Arthur Burnand
Henry Baring
G. H. L. Boynton
Wm. D. Seymour
Wm. W. King
John Gibsone
Cornets.—James Duncan
Walter R. Nolan
Henry Marshall
George Cleghorn
Hon. W. H. Curzon
Charles Waymouth
Robert Bainbridge
Paymaster.—John Stephenson
Adjutant.—John Chadwick
Quartermaster.—Dennis O'Hara
Surgeon.—H. H. Massey, M.D.
Asst.-Surgeon.—St. John Stanley
Vet. Surgeon.—Wm. Partridge

Appendix A

1857

Colonel.—Sir J. M. Wallace, K.H.
Lieut.-Colonel.—H. R. Benson
Major.—A. Learmonth
Captains.—W. Morris, C.B. (Major)
R. White
Sir W. Gordon, Bart.
L. E. Knight
J. Macartney
W. J. P. Watson
Lieutenants.—T. Taylor, R.M.
A. Burnand
H. Baring
G. H. L. Boynton
W. D. Seymour
W. W. King
J. Gibsone
Cornets.—J. Duncan
W. R. Nolan
H. Marshall
G. Cleghorn
Hon. W. H. Curzon
C. Waymouth
R. Bainbridge
Paymaster.—J. Stephenson
Adjutant.—J. Duncan
Quartermaster.—W. Garland
Surgeon.—H. H. Massey, M.D.
Asst.-Surgeon.—St. John Stanley
Vet. Surgeon.—W. Partridge

1858

Colonel.—Sir J. M. Wallace, K.H.
Lieut.-Colonels.—H. R. Benson
J. R. H. Rose
Majors.—A. Learmonth
W. Morris, C.B. (Lt.-Col.)
Captains.—R. White
Sir W. Gordon, Bart.
L. E. Knight
J. Macartney
A. Burnand
Sir G. H. Leith, Bart.
D. C. Lowe
Captains.—T. Taylor
H. Baring
H. A. Sarel
Lieutenants.—W. D. Seymour
W. W. King
J. Gibsone
J. Duncan
W. R. Nolan
H. Marshall
Hon. H. W. Curzon
C. Waymouth
R. Bainbridge
H. E. Wood
T. Gonne
Cornets.—A. Gooch
F. J. King
J. Harding
R. D. Macgregor
J. G. Scott
W. S. Tucker
R. T. Goldsworthy
J. T. Fraser
H. W. F. Harrison
E. A. Corbet
Paymaster.—F. L. Bennett
Adjutant.—J. Duncan
Quartermaster.—W. Garland
Surgeon.—E. Mockler
Asst.-Surgeons.—G. C. Clery
Y. H. Johnson
Vet. Surgeon.—W. Partridge

1859

Colonel.—Sir J. M. Wallace, K.H.
Lieut.-Colonels.—H. R. Benson
J. R. H. Rose
Majors.—A. Learmonth
R. White
Captains.—Sir W. Gordon, Bart.
L. E. Knight
J. Macartney
Sir G. H. Leith, Bart.
D. C. Lowe
T. Taylor

Captains.—H. Baring
H. A. Sarel
C. Steel
W. D. Seymour
Lieutenants.—J. Gibsone
J. Duncan
W. R. Nolan
H. Marshall
Hon. W. H. Curzon
C. Waymouth
R. Bainbridge
H. E. Wood, V.C.
T. Gonne
F. J. King
J. Harding
Cornets.—R. D. Macgregor
J. G. Scott
W. S. Tucker
J. I. Fraser
R. T. Goldsworthy
H. W. F. Harrison
E. A. Corbet
Paymaster.—G. B. Belcher
Adjutant.—J. Duncan
Quartermaster.—Wm. Garland
Riding-Master.—G. Pumfrett
Surgeon.—J. Kellie, M.D.
Asst.-Surgeons.—Y. H. Johnson
G. C. Clery
Vet. Surgeon.—W. Partridge

1860

Colonel.—Sir J. M. Wallace, K.H.
Lieut.-Colonels.—H. R. Benson
A. Learmonth
Majors.—R. White
Sir. W. Gordon, Bart.
Captains.—L. E. Knight
J. Macartney
Sir G. H. Leith
D. C. Lowe
H. A. Sarel
C. Steel
W. R. Nolan

Captains.—J. Gibsone
H. Marshall
Lieutenants.—J. Duncan
Hon. W. H. Curzon
C. Waymouth
R. Bainbridge
H. E. Wood, V.C.
T. Gonne
J. Harding
A. J. Billing
R. D. Macgregor
J. G. Scott
R. T. Goldsworthy
Cornets.—J. I. Fraser
H. W. F. Harrison
H. R. Abadie
G. J. B. Bruce
H. W. Young
G. Rosser
F. W. Blumberg
Paymaster.—G. B. Belcher
Adjutant.—J. Duncan
Quartermaster.—W. Garland
Riding-Master.—G. Pumfrett
Surgeon.—G. Kellie, M.D.
Asst-Surgeons.—Y. H. Johnson
G. C. Clery
Veterinary Surgeon.—J. Ferris

1861

Colonel.—Sir J. M. Wallace, K.H.
Lt.-Col. & Col.—H. R. Benson, C.B.
Lieut.-Colonel.—Robert White
Lt.-Col. & Col.—J. C. H. Gibsone
Majors.—Sir W. Gordon, Bart.
L. E. Knight
Captains.—John Macartney
D. C. Lowe
H. A. Sarel
W. R. Nolan
John Gibsone
James Duncan
Hon. W. H. Curzon
Charles Waymouth

Captains.—James Goldie
Robert Bainbridge
Lieutenants.—H. E. Wood, V.C.
T. Gonne
J. Harding.
A. J. Billing
R. D. Macgregor
J. G. Scott
R. T. Goldsworthy
J. I. Fraser
H. W. F. Harrison
H. R. Abadie
Cornets.—G. J. B. Bruce
H. W. Young
George Rosser
F. W. Blumberg
George Pumfrett
H. A. Robinson
J. D. Jackson
Edward Corbett
E. H. Maunsell
Paymaster.—De P. O'Kelly
Adjutant.—G. Pumfrett
Riding-Master.—Thomas Martin
Quartermaster.—W. Garland
Surgeon.—James Kellie, M.D.
Asst.-Surgeons.—Sam. Fuller
DavidCullen,M.D.
Veterinary Surgeon.—J. Ferris

1862

Colonel.—Sir J. M. Wallace, K.H.
Lt.-Col. & Col.—H. R. Benson
Lieut.-Colonel.—Robert White
Lt.-Col. & Col.—J. C. H. Gibsone
Majors.—Sir W. Gordon, Bart.
L. E. Knight
Captains.—D. C. Lowe
H. A. Sarel (B.Lt.-Col.)
W. R. Nolan
John Gibsone
James Duncan
Hon. W. H. Curzon
Charles Waymouth

Captains.—James Goldie
Robert Bainbridge
H. E. Wood, V.C.
Lieutenants.—T. Gonne
James Harding
A. J. Billing
R. T. Goldsworthy
H. R. Abadie
B. Chamley
G. J. B. Bruce
H. W. Young
George Rosser
Cornets.—F. W. Blumberg
George Pumfrett
H. A. Robinson
T. D. Jackson
Edward Corbett
E. H. Maunsell
E. W. Pritchard
S. Y. Clark
H. Faulkner
Harris St. J. Dick
Adjutant.—George Pumfrett
Paymaster.—De P. O'Kelly
Riding-Master.—Thomas Martin
Quartermaster.—William Garland
Surgeon.—James Kellie, M.D.
Asst.-Surgeons.—Sam. Fuller
D. Cullen, M.D.
Veterinary-Surgeon.—J. Ferris

1863

Colonel.—Sir J. M. Wallace, K.H.
Lieut.-Colonels.—Robert White
Sir W. Gordon, Bt.
Majors.—L. E. Knight
Drury C. Lowe
Captains.—H. A. Sarel (B.Lt.-Col.)
Walter R. Nolan
James Duncan
Hon. W. H. Curzon
C. Waymonth
James Goldie
Robert Bainbridge

Captions.—T. Gonne
T. W. S. Miles
W. Balfe
Lieutenants.—A. J. Billing
R. T. Goldsworthy
H. R. Abadie
B. Chamley
H. W. Young
George Rosser
F. W. Blumberg
G. Pumfrett
H. A. Robinson
W. S. Browne
Cornets.—J. D. Jackson
E. Corbett
E. H. Maunsell
E. W. Pritchard
S. Y. Clark
H. Faulkner
H. St. J. Dick
Robert Blair
J. C. Symonds
Paymaster.—De P. O'Kelly
Adjutant.—G. Pumfrett
Riding-Master.—Thomas Martin
Quartermaster.—W. Garland
Surgeon.—J. Kellie, M.D.
Asst-Surgeons.—Sam. Fuller
David Cullen, M.D.
Veterinary Surgeon.—John Ferris

1864

Colonel.—Sir J. M. Wallace, K.H.
Lieut.-Colonels.—Robert White
Sir W. Gordon, Bt.
Majors.—L. E. Knight
Drury C. Lowe
Captains.—H. A. Sarel (Lieut.-Col.)
W. R. Nolan
James Duncan
Hon. W. H. Curzon
C. Waymouth
J. Goldie
Robert Bainbridge
Thomas Gonne
T. W. S. Miles
W. Balfe
Lieutenants.—A. J. Billing
R. T. Goldsworthy
H. R. Abadie
B. Chamley
H. W. Young
George Rosser
F. W. Blumberg
George Pumfrett
H. A. Robinson
Cornets.—J. D. Jackson
E. Corbett
E. H. Maunsell
S. Y. Clark
H. Faulkner
H. St. J. Dick
Robert Blair
J. C. Symonds
W. A. Ellis
Paymaster.—De P. O'Kelly
Adjutant.—George Pumfrett
Riding-Master.—T. Martin
Quartermaster.—W. Garland
Surgeon.—J. Kellie, M.D.
Asst.-Surgeons.—J. Fuller
D. Cullen, M.D.
Vet. Surgeon.—James Lambert

1865

Colonel.—Sir J. M. Wallace, K.H.
Lieut.-Colonel.—Robert White
L. E. Knight
Majors.—Drury C. Lowe
Hon. W. H. Curzon
Captains.—H. A. Sarel (B. Lt.-Col.)
W. R. Nolan
James Duncan
C. Waymouth
J. Goldie
R. Bainbridge
T. Gonne
T. W. S. Miles

Lieutenants.—A. J. Billing
R. T. Goldsworthy
H. R. Abadie
H. W. Young
George Rosser
F. W. Blumberg
George Pumfrett
H. A. Robinson
J. D. Jackson
Edward Corbett
Cornets.—E. H. Maunsell
S. Y. Clark
H. Faulkner
J. C. Symonds
William A. Ellis
H. T. S. Carter
William Watt
H. Bancroft
Paymaster.—De P. O'Kelly
Adjutant.—George Pumfrett
Riding-Master.—T. Martin
Quartermaster.—J. Berryman, V.C.
Surgeon.—James Kellie, M.D.
Asst-Surgeon.—S. A. Lithgow
Veterinary Surgeon.—J. Lambert

1866

Colonel.—Sir J. M. Wallace, K.H.
Lieut.-Colonel.—Robert White
Majors.—Drury C. Lowe
Hon. W. H. Curzon
Captains.—H. A. Sarel (B. Lt.-Col.)
W. R. Nolan
Charles Waymouth
Robert Bainbridge
T. Gonne
William A. Battine
Sir John Hill, Bart.
George C. Robinson
Lieutenants.—Arthur J. Billing
Henry R. Abadie
H. W. Young
F. W. Blumberg
George Pumfrett
Lieutenants.—H. A. Robinson
Edward Corbett
W. G. Walmesley
E. H. Maunsell
Cornets.—S. Y. Clark
H. Faulkner
John C. Symonds
Harry T. S. Carter
H. Bancroft
E. B. Callander
S. M. Benson
W. Brougham
Paymaster.—De P. O'Kelly
Adjutant.—George Pumfrett
Riding-Master.—Thomas Martin
Quartermaster.—J. Berryman, V.C.
Surgeon.—James Kellie, M.D.
Asst-Surgeon.—S. A. Lithgow
Vet. Surgeon.—James Lambert

1867

Colonel.—Sir J. M. Wallace, K.H.
Lieut.-Colonel.—Drury C. D. Lowe
Major.—Hon. W. H. Curzon
Major Lieut.-Col.—Henry A. Sarel
Captains.—Walter R. Nolan
Charles Waymouth
Robert Bainbridge
T. Gonne
Sir J. Hill, Bt. (B. Maj.)
George C. Robinson
Sam. Boulderson
W. A. Battine
Lieutenants.—Henry R. Abadie
F. W. Blumberg
H. A. Robinson
W. G. Walmesley
Stanley Y. Clark
H. Bancroft
Thomas A. Cooke
Hon. A. W. Erskine
Cornets.—E. B. Callander
S. M. Benson
W. Brougham

Cornets.—Thomas Crowe
E. V. W. Edgell
Sir Charles Nugent, Bart.
C. W. J. Unthank
Ernest A. Belford
Paymaster.—De P. O'Kelly
Adjutant.—A. J. Billing
Riding-Master.—Thomas Martin
Quartermaster.—John Berryman, V.C.
Surgeon.—James Kellie, M.D.
Asst.-Surgeon.—S. A. Lithgow
Vet. Surgeon.—James Lambert

1868

Colonel.—C. W. M. Balders, C.B.
Lieut.-Colonel.—Drury C. Lowe
Majors.—Hon. W. H. Curzon
H. A. Sarel (B. Lt.-Col.)
Captains.—W. R. Nolan
Charles Waymouth
Robert Bainbridge
T. Gonne
W. A. Battine
G. C. Robinson
S. Boulderson
F. W. Blumberg
Lieutenants.—H. A. Robinson
W. G. Walmesley
S. Y. Clark
Thomas A. Cooke
Hon. A. W. Erskine
S. M. Benson
W. Brougham
Thomas Crowe
G. H. L. Pellew
Cornets.—E. V. W. Edgell
Sir Charles Nugent, Bart.
C. W. J. Unthank
Ernest A. Belford
James F. Alexander
Hon. J. P. Bouverie
John Brown
William Bashford
Paymaster.—De P. O'Kelly (Hon. Captain)
Adjutant.—John Brown
Riding-Master.—Thomas Martin
Quartermaster.—J. Berryman, V.C.
Surgeon.—Arthur Greer
Asst.-Surgeon.—J. E. O'Loughlin
Vet. Surgeon.—James Lambert

1869

Colonel.—C. W. M. Balders, C.B.
Lieut.-Colonel.—Drury C. D. Lowe
Majors.—Hon. W. H. Curzon
Henry A. Sarel (Lt.-Col.)
Captains.—W. R. Nolan
Charles Waymouth
Robert Bainbridge
T. Gonne
G. C. Robinson
Samuel Boulderson
F. W. Blumberg
H. A. Robinson
Lieutenants.—W. G. Walmesley
S. Y. Clark
T. A. Cooke
S. M. Benson
Thomas Crowe
G. H. L. Pellew
Sir C. Nugent, Bart.
C. W. J. Unthank
Cornets.—Ernest A. Belford
J. F. Alexander
Hon. J. P. Bouverie
John Brown (Adj.)
William Bashford
W. T. S. Kevill-Davies
C. E. Swaine
R. N. Humble
Paymaster.—De P. O'Kelly, (Hon. Captain)
Riding-Master.—Thomas Martin
Quartermaster.—J. Berryman, V.C.
Surgeon.—A. J. Greer
Asst.-Surgeon.—J. E. O'Loughlin

Vet. Surgeon.—James Lambert

1870

Colonel.—C. W. M. Balders, C.B.
Lieut.-Colonel.—Drury C. Lowe
Majors.—Hon. W. H. Curzon
W. R. Nolan
Captains.—Charles Waymouth
Robert Bainbridge
T. Gonne
G. C. Robinson
S. Boulderson
F. W. Blumberg
S. Y. Clark
J. C. Duke
Lieutenants.—T. A. Cooke
S. M. Benson
Thomas Crowe
E. V. W. Edgell
C. W. J. Unthank
Ernest A. Belford
J. F. Alexander
Hon. J. P. Bouverie
William Bashford
Cornets.—W. T. S. Kevill-Davies
Charles E. Swaine
R. N. Humble
Charles E. Arkwright
Paymaster.—De P. O'Kelly
Adjutant.—J. Brown (Lieut.)
Riding-Master.—R. H. Boyle
Surgeon.—A. J. Greer
Quartermaster.—J. Berryman, V.C.
Asst.-Surgeon.—J. E. O'Loughlin
Vet. Surgeon.—James Lambert

1871

Colonel.—C. W. M. Balders, C.B.
Lieut.-Colonel.—D. C. Drury Lowe
Majors.—W. R. Nolan
Robert Bainbridge
Captains.—T. Gonne
G. C. Robertson
S. Boulderson
F. W. Blumberg
S. Y. Clark
J. C. Duke
Thomas A. Cooke
S. M. Benson
Lieutenants.—E. V. W. Edgell
C. W. J. Unthank
E. A. Belford
J. F. Alexander
Hon. J. P. Bouverie
John Brown (Adj.)
William Bashford
W. T. S. Kevill-Davies
C. E. Swaine
R. N. Humble
Cornets.—C. E. Arkwright
Thomas Mack
A. E. De Butts
Paymaster.—De P. O'Kelly
Riding-Master.—R. H. Boyle
Quartermaster.—J. Berryman, V.C.
Surgeon.—A. J. Greer
Asst.-Surgeon.—Ed. Hoile, M.D.
Veterinary Surgeon.—J. Lambert

1872

Colonel.—C. W. M. Balders, C.B.
(Lieut.-General)
Lt.-Col.—D. C. Drury Lowe (Col.)
Majors.—W. R. Nolan
G. C. Robertson
Captains.—T. Gonne
S. Boulderson
F. W. Blumberg
S. Y. Clark
J. C. Duke
Thomas A. Cooke
S. M. Benson
C. W. J. Unthank
Lieutenants.—E. V. W. Edgell
E. A. Belford
J. F. Alexander
Hon. J. P. Bouverie

Lieutenants.—John Brown (Adj.)
W. T. S. Kevill-Davies
Charles E. Swaine
Robert N. Humble
H. M. Barton
C. E. Arkwright
Sub-Lieutenants.—T. Mack
A. E. de Butts
G. A. Wood
Paymaster.—J. W. Smith
Riding-Master.—J. Berryman, V.C.
Surgeon.—Arthur J. Greer
Assistant-Surgeon.—E. Hoile, M.D.
Veterinary Surgeon.—J. Lambert

1873

Colonel.—C. W. M. Balders, C.B. (Lieut.-General)
Lt.-Col.—D. C. Drury Lowe (Col.)
Majors.—W. R. Nolan
G. C. Robertson
Captains.—Thomas Gonne
Samuel Boulderson
F. W. Blumberg
S. Y. Clark
J. C. Duke
T. A. Cooke
S. M. Benson
C. W. J. Unthank
Lieutenants.—E. V. W. Edgell
E. A. Belford
J. F. Alexander
Hon. J. P. Bouverie
John Brown (Adj.)
W. T. S. Kevill-Davies
Charles E. Swaine
R. N. Humble
C. E. Arkwright
Thomas Mack
Sub-Lieutenants.—George A. Wood
Percy Wormald
John M. Russell
Paymaster.—John W. Smith
Riding-Master.—Richard H. Boyle
Quartermaster.—J. Berryman, V.C.
Surgeon.—Arthur Greer
Assistant-Surgeon.—E. Hoile, M.D.
Veterinary Surgeon.—J. Lambert

1874

Colonel.—C. W. M. Balders, C.B. (Lieut.-General)
Lt.-Col.—D. C. Drury Lowe (Col.)
Major.—Walter R. Nolan
Captains.—Thomas Gonne
Samuel Boulderson
Frederick W. Blumberg
S. Y. Clark
J. C. Duke
Thomas A. Cooke
S. M. Benson
E. V. W. Edgell
Lieutenants.—Ernest A. Belford
J. F. Alexander
Hon. J. P. Bouverie
John Brown (Adj.)
W. T. S. Kevill-Davies
Charles E. Swaine
Robert N. Humble
C. E. Arkwright
Thomas Mack
George A. Wood
Mortimer G. Neeld
Sub-Lieutenants.—Percy Wormald
John M. Russell
C. H. Purvis
Paymaster.—J. W. Smith
Riding-Master.—Richard Boyle
Quartermaster.—J. Berryman, V.C.
Medical Officer.—Arthur J. Greer
Veterinary Surgeon.—J. Lambert

1875

Colonel.—C. W. M. Balders, C.B. (Lieut.-General)

Appendix A

Lt-Col.—D. C. Drury Lowe (Col.)
Major.—Thomas Gonne
Captains.—Samuel Boulderson
F. W. Blumberg
S. Y. Clark
S. M. Benson
E. V. W. Edgell
Ernest A. Belford
Lieutenants.—James F. Alexander
Hon. J. P. Bouverie
John Brown (Adj.)
W. T. S. Kevill-Davies
Charles E. Swaine
Charles E. Arkwright
Thomas Mack
Percy Wormald
John M. Russell
George A. Wood
Mortimer G. Neeld
H. C. Jenkins
Sub-Lieutenant.—C. H. Purvis
Riding-Master.—Richard H. Boyle
Quartermaster.—J. Berryman, V.C.
Medical Officer.—A. C. McTavish
Veterinary Surgeon.—J. Lambert

1876

Colonel.—J. C. Hope Gibsone (Lieut.-General)
Lt.-Col.—D. C. Drury Lowe (Col.)
Major.—Thomas Gonne
Captains.—Samuel Boulderson
F. W. Blumberg
S. Y. Clark
J. C. Duke
Thomas A. Cooke
S. M. Benson
E. V. Wyatt-Edgell
Ernest A. Belford
Lieutenants.—J. F. Alexander
Hon J. P. Bouverie
John Brown (Adj.)
W.T.S.Kevill-Davies
Lieutenants.—Charles E. Swaine
Charles E. Arkwright
Thomas Mack
Percy Wormald
John M. Russell
George A. Wood
M. G. Neeld
H. C. Jenkins
C. H. Purvis
Sub-Lieut.—C. F. S. Anstruther-Thomson
Riding-Master.—Richard H. Boyle
Quartermaster.—J. Berryman, V.C.
Surgeon-Major.—A. C. McTavish
Veterinary Surgeon.—J. Lambert

1877

Colonel-in-Chief
H.R.H. Duke of Cambridge, Field Marshal, Commanding-in-Chief
Colonel.—J. C. Hope Gibsone (Lieut.-General)
Lt.-Col.—D. C. Drury Lowe (Col.)
Majors.—Thomas Gonne
Samuel Boulderson
Captains.—Fred. W. Blumberg
S. Y. Clark
J. C. Duke
Thomas A. Cooke
S. M. Benson
E. V. Wyatt Edgell
Ernest A. Belford
James F. Alexander
Lieutenants.—Hon. J. P. Bouverie
John Brown (Adj.)
W. T. S. Kevill-Davies
Charles E. Swaine
Charles E. Arkwright
Percy Wormald
John M. Russell
George A. Wood
M. G. Neeld

Lieutenants.—H. C. Jenkins
C. H. Purvis
H. Fortescue
Riding-Master.—R. H. Boyle
Quartermaster.—J. Berryman, V.C.
Surgeon-Major.—A. C. McTavish
Vet. Surgeon.—James Lambert

1878

Colonel-in-Chief.
H.R.H. Duke of Cambridge, Field Marshal, Commanding-in-Chief
Colonel.—J. C. Hope Gibsone (Gen.)
Lieutenant-Colonel.—D. C. Drury Lowe (Col.)
Majors.—Thomas Gonne
S. Boulderson
Captains.—S. Y. Clark
J. C. Duke
T. A. Cooke
S. M. Benson
E. V. Wyatt-Edgell
Ernest A. Belford
J. F. Alexander
Hon. J. P. Bouverie
Lieutenants.—John Brown (Adj.)
W. T. S. Kevill-Davies
C. E. Swaine
J. M. Russell
G. A. Wood
M. G. Neeld
H. C. Jenkins
C. H. Purvis
H. Fortescue
Sub-Lts.—F. J. C. Frith
T. A. Steele
E. B. Herbert
Hon. L. H. D. Fortescue
Riding-Master.—R. H. Boyle
Quartermaster.—J. Berryman, V.C.
Vet. Surgeon.—James Lambert

1879

Colonel-in-Chief.
H.R.H. Duke of Cambridge, Field Marshal, Commanding-in-Chief
Colonel.—J. C. Hope Gibsone (Gen.)
Lieut.-Colonel.—Thomas Gonne
Major.—Samuel Boulderson
Captains.—S. Y. Clark
James C. Duke
Thomas A. Cooke
S. M. Benson
E. V. Wyatt Edgell
E. A. Belford
James F. Alexander
Hon. J. P. Bouverie
Lieuts.—John Brown (Adj.)
W. T. S. Kevill-Davies
C. E. Swaine
J. M. Russell
George A. Wood
M. G. Neeld
H. C. Jenkins
C. H. Purvis
F. J. Cockayne Frith
Henry Fortescue
Thomas A. Steele
E. B. Herbert
Hon. L. H. D. Fortescue
2nd Lieuts.—C. J. Anstruther Thomson
C. H. Butler
F. D. H. St. Quintin
Riding-Master.—R. H. Boyle
Quartermaster.—J. Berryman, V.C.
Vet. Surgeon.—James Lambert

1880

Colonel-in-Chief.
H.R.H. Duke of Cambridge, Field Marshal, Commanding-in-Chief
Colonel.—J. C. Hope Gibsone (Gen.)
Lieut.-Colonel.—Thomas Gonne

Appendix A

Major.—Samuel Boulderson
Captains.—S. Y. Clark
J. C. Duke
Thomas A. Cooke
S. M. Benson
Captains.—Ernest A. Belford
James F. Alexander
Hon. J. P. Bouverie
W. T. S. Kevill-Davies
Lieutenants.—Charles E. Swaine
John M. Russell
Geo. A. Wood
M. G. Neeld
H. C. Jenkins
C. H. Purvis
H. Fortescue
Thos. A. Steele
E. B. Herbert
Hon. L. H. D. Fortescue
2nd Lieuts.—C. J. Anstruther Thomson
Chas. H. Butler
F. D. H. St. Quintin
W. G. Renton
M. H. Woods
James H. Dyer
Paymaster.—J. Brown (Hon. Cap.)
Adj.—Hon. L. H. D. Fortescue
Riding-Master.—John Perry
Quartermaster.—J. Berryman, V.C.
Vet. Surgeon.—James Lambert

1881

Colonel-in-Chief.
H.R.H. Duke of Cambridge, Field Marshal, Commanding-in-Chief
Colonel.—J. C. Hope Gibsone (Gen.)
Lieut.-Colonel.—Thos. Gonne
Major.—Samuel Boulderson
Captains.—S. Y. Clark
J. C. Duke
Thos. A. Cooke
Captains.—S. M. Benson
Ernest A. Belford
Hon. J. P. Bouverie
John M. Russell
Lieutenants.—Geo. A. Wood
M. G. Neeld
H. C. Jenkins
C. H. Purvis
Henry Fortescue
Thos. A. Steele
E. B. Herbert
Hon. L. H. D. Fortescue (Adj.)
2nd Lieuts.—C. J. Anstruther Thomson
Chas. H. Butler
W. G. Renton
J. H. Dyer
C. Coventry
Paymaster.—J. Brown (Hon. Capt.)
Riding-Master.—John Perry
Quartermaster.—Douglas Shawe

1882

Colonel-in-Chief.
H. R. H. Duke of Cambridge, Field Marshal, Commanding-in-Chief
Colonel.—J. C. Hope Gibsone (Gen.)
Lieut.-Colonels.—Samuel Boulderson
S. Y. Clark
Majors.—J. C. Duke
Thos. A. Cooke
S. M. Benson.
Captains.—Ernest A. Belford
Hon. J. P. Bouverie
John M. Russell
F. W. Benson
Lieutenants.—M. G. Neeld
H. C. Jenkins
Chas. H. Purvis
Henry Fortescue
Thos. A. Steele
E. B. Herbert

Lieutenants.—Hon. L. H. D. Fortescue (Adj.)
C. J. Anstruther Thomson
Chas. H. Butler
W. G. Renton
James H. Dyer
Chas. Coventry
Thos. H. Standbridge
Paymaster.—John Brown (Hon. Capt.)
Riding-Master.—John Perry
Quartermaster.—Douglas Shawe

1883

Colonel-in-Chief.
H. R. H. Duke of Cambridge, Field Marshal, Commanding-in-Chief
Colonel.—J. C. Hope Gibsone (Gen.)
Lieut.-Colonels.—Sam. Boulderson
Thos. A. Cooke
Majors.—S. M. Benson
Ernest A. Belford
Hon. J. P. Bouverie
Captains.—F. W. Benson
M. G. Neeld
H. C. Jenkins
C. H. Purvis
Henry Fortescue
Lieutenants.—Thos. A. Steele
E. B. Herbert
Hon. L. H. D. Fortescue (Adj.)
C. J. Anstruther Thomson
Chas. H. Butler
Wm. G. Renton
James H. Dyer
Chas. Coventry
T. H. Standbridge
H. W. R. Ricardo
Hon. H. A. Lawrence
G. C. C. D'Aguilar
Paymaster.—J. M. Russell (H. Capt.)
Riding-Master.—John Perry
Quartermaster.—Douglas Shawe

1884

Colonel-in-Chief.
H. R. H. Duke of Cambridge, Field Marshal, Commanding-in-Chief
Colonel.—J. C. Hope Gibsone (Gen.)
Lieut.-Colonels.—Sam. Boulderson
Thos. A. Cooke
Majors.—S. M. Benson
Ernest A. Belford
Hon. J. P. Bouverie
Captains.—F. W. Benson
M. G. Neeld
H. C. Jenkins
C. . Purvis
Henry Fortescue
Lieutenants.—Thomas A. Steele
E. B. Herbert
Hon. L. H. D. Fortescue (Adj.)
C. J. Anstruther Thomson
Chas. H. Butler
Wm. G. Renton
James H. Dyer
Chas. Coventry
T. H. Standbridge
H. W. R. Ricardo
Hon. H. A. Lawrence
G. C. C. D'Aguilar
Paymaster.—J. M. Russell (H. Capt.)
Riding-Master.—John Perry
Quartermaster.—Douglas Shawe

1885

Colonel-in-Chief.
H. R. H. Duke of Cambridge, Field Marshal, Commanding-in-Chief
Colonel.—H. R. Benson, C.B. (Gen.)

Appendix A

Lieut.-Colonels.—S. Boulderson
Thos. A. Cooke
Majors.—S. M. Benson
E. A. Belford
Hon. J. P. Bouverie
Captains.—F. W. Benson
M. G. Neeld
H. C. Jenkins
C. H. Purvis
H. Fortescue
T. A. Steele
Lieutenants.—E. B. Herbert
Hon. L. H. D. Fortescue
C. J. Anstruther Thomson
C. H. Butler
W. G. Renton
J. H. Dyer
C. Coventry
T. H. Standbridge
H. W. R. Ricardo
Hon. H. A. Lawrence
G. C. C. D'Aguilar
G. F. Milner
C. A. S. Warner
Paymaster.—J. M. Russell (Hon. Captain)
Adjutant.—Hon. L. H. D. Fortescue
Riding-Master.—H. M'Gee
Quartermaster.—D. Shawe

1886

Colonel-in-Chief.
H. R. H. Duke of Cambridge, Field Marshall, Commanding-in-Chief
Colonel.—H. R. Benson, C.B. (Gen.)
Lieut.-Colonels.—S. Boulderson
T. A. Cooke
Majors.—S. M. Benson
E. A. Belford
Hon. J. P. Bouverie
Captains.—F. W. Benson
M. G. Neeld
H. C. Jenkins
C. H. Purvis
H. Fortescue
T. A. Steele
Lieutenants.—E. B. Herbert
Hon. L. H. D. Fortescue
C. J. Anstruther Thomson
C. H. Butler
W. G. Renton
J. H. Dyer
C. Coventry
T. H. Standbridge
H. W. R. Ricardo
Hon. H. A. Lawrence
G. C. C. D'Aguilar
G. F. Milner
C. A. S. Warner
B. P. Portal
Paymaster.—J. M. Russell (Hon. Captain)
Adjutant.—C. Coventry
Riding-Master.—H. M'Gee (Hon. Captain)
Quartermaster.—D. Shawe (Hon. Captain)

1887

Colonel-in-Chief.
H. R. H. Duke of Cambridge, Field Marshall, Commanding-in-Chief
Colonel.—H. R. Benson, C.B. (Gen.)
Lieut.-Colonels.—T. A. Cooke
S. M. Benson
Majors.—E. A. Belford
Hon. J. P. Bouverie
F. W. Benson
M. G. Neeld
H. C. Jenkins
Captains.—C. H. Purvis

Captains.—H. Fortescue
T. A. Steele
E. B. Herbert
Hon. L. H. D. Fortescue
C. J. Anstruther Thomson

Lieutenants.—C. H. Butler
W. G. Renton
C. Coventry
H. W. R. Ricardo
Hon. H. A. Lawrence
G. C. C. D'Aguilar
G. F. Milner
E. W. N. Pedder
C. A. S. Warner
B. P. Portal
A. J. T., Viscount Clandeboye
A. Rawlinson
N. T. Nickalls
E. D. Miller
H. M. Jessel
V. S. Sandeman

Paymaster.—J. M. Russell (Hon. Captain)
Adjutant.—C. Coventry (Lieut.)
Riding-Master.—H. M'Gee
Quartermaster.—D. Shawe

1888

Colonel-in-Chief.

H. R. H. Duke of Cambridge, Field Marshal, Commander-in-Chief

Colonel.—H. R. Benson, C.B. (Gen.)
Lieut.-Colonels.—T. A. Cooke
S. M. Benson

Majors.—E. A. Belford
Hon. J. P. Bouverie
F. W. Benson
M. G. Neeld
H. C. Jenkins

Captains.—C. H. Purvis
H. Fortescue
J. A. Steele
E. B. Herbert
Hon. L. H. D. Fortescue
C. J. Anstruther Thomson
C. H. Butler

Lieutenants.—W. G. Renton
C. Coventry
H. W. R. Ricardo
Hon. H. A. Lawrence
G. C. C. D'Aguilar
G. F. Milner
E. W. N. Pedder
C. A. S. Warner
B. P. Portal
A. J. T., Viscount Clandeboye
N. T. Nickalls
E. D. Miller
H. M. Jessel
V. S. Sandeman

2nd Lieuts.—R. du P. Grenfell
T. G. Collins

Paymaster.—J. M. Russell (Capt.)
Adjutant.—C. Coventry
Riding-Master.—H. M'Gee
Quartermaster.—D. Shawe

1889

Colonel-in-Chief.

H. R. H. Duke of Cambridge, Field Marshal, Commander-in-Chief

Colonel.—H. R. Benson, C.B. (Gen.)
Lieut.-Colonel.—S. M. Benson

Majors.—E. A. Belford
Hon. J. P. Bouverie
F. W. Benson
M. G. Neeld
H. C. Jenkins

Captains.—C. H. Purvis
H. Fortescue
T. A. Steele
E. B. Herbert

Captains.—Hon. L. H. D. Fortescue
C. J. Anstruther
W. G. Renton
C. Coventry (Adjutant)
H. W. R. Ricardo
Lieutenants.—Hon. H. A. Lawrence
G. C. C. D'Aguilar
G. F. Milner
C. A. S. Warner
F. P. M. Maryon-Wilson
B. P. Portal
A. J. T., Earl of Ava
A. Rawlinson
N. T. Nickalls
E. D. Miller
H. M. Jessel
V. S. Sandeman
2nd Lieuts.—R. du P. Grenfell
T. G. Collins
Prince Adolphus of Teck
H. C. Noel
Paymaster.—J. M. Russell
Riding-Master.—H. M'Gee
Quartermaster.—D. Shawe

1890.

Colonel-in-Chief.
H.R.H. Duke of Cambridge, Field Marshal, Commander-in-Chief
Colonel.—H .R. Benson, C.B. (Gen.)
Lieut.-Colonel.—S. M. Benson
Majors.—E. A. Belford
Hon. J. P. Bouverie
F. W. Benson
M. G. Neeld
H. C. Jenkins
Captains.—C. H. Purvis
H. Fortescue
T. A. Steele
E. B. Herbert
Hon. L .H. D. Fortescue
C. J. Anstruther
W. G. Renton
C. Coventry
H. W. R. Ricardo
Lieutenants.—Hon. H. A. Lawrence
G. C. C. D'Aguilar
G. F. Milner
C. A. S. Warner
F. P. M. Maryon-Wilson
B. P. Portal
A. J. T., Earl of Ava
A. Rawlinson
N. T. Nickalls
E. D. Miller
H. M. Jessel
V. S. Sandeman
2nd Lieuts.—T. G. Collins
Prince Adolphus of Teck
H. C. Noel
W. F. Egerton
W. A. Tilney
Paymaster.—J. M. Russell
Adjutant.—C. Coventry
Riding-Master.—H. M'Gee
Quartermaster.—D. Shawe

1891.

Colonel-in-Chief.
H.R.H. Duke of Cambridge, Field Marshal, Commander-in-Chief
Col.—H. R. Benson, C.B (Gen.)
Lieutenant-Colonel.—S. M. Benson
Majors.—E. A. Belford
Hon. J. P. Bouverie
F. W. Benson
M. G. Neeld
H. C. Jenkins
Captains.—C. H. Purvis
H. Fortescue
T. A. Steele
E. B. Herbert

Captains.—Hon. L. H. D. Fortescue
C. J. Anstruther
W. G. Renton
C. Coventry
H. W. R. Ricardo
Lieutenants.—Hon. H. A. Lawrence
G. C. C. D'Aguilar
G. F. Milner
C. A. S. Warner
F. P. M. Maryon-Wilson
B. P. Portal
A. J. T., Earl of Ava
A. Rawlinson
N. T. Nickalls
E. D. Miller
H. M. Jessel
V. S. Sandeman
2nd Lieuts.—T. G. Collins
Prince Adolphus of Teck
H. C. Noel
W. F. Egerton
W. A. Tilney
Adjutant.—Hon. H. A. Lawrence
Riding-Master.—H. M'Gee
Quartermaster.—D. Shawe

Captains.—C. J. Anstruther
W. G. Renton
C. Coventry
H. W. R. Ricardo
Lieutenants.—Hon. H. A. Lawrence
G. C. C. D'Aguilar
G. F. Milner
C. A. S. Warner
F. P. M. Maryon-Wilson
B. P. Portal
A. J. T., Earl of Ava
A. Rawlinson
N. T. Nickalls
E. D. Miller
H. M. Jessel
V. S. Sandeman
2nd Lieuts.—T. G. Collins
Prince Adolphus of Teck
H. C. Noel
W. F. Egerton
W. A. Tilney
Adjt.—Hon. H. A. Lawrence
Riding-Master.—W. Pilley (Hon. Lieutenant)
Quartermaster.—D. Shawe

1892.

Colonel-in-Chief.
H.R.H. Duke of Cambridge, Field Marshal, Commander-in-Chief
Colonel.—H. R. Benson (Gen.)
Lieutenant-Colonel.—S. M. Benson
Majors.—E. A. Belford
Hon. J. P. Bouverie
F. W. Benson
M. G. Neeld
H. C. Jenkins
Captains.—C. H. Purvis
H. Fortescue
E. B. Herbert
Hon. L. H. D. Fortescue

1893.

Colonel-in-Chief.
H.R.H. Duke of Cambridge, Field Marshal, Commander-in-Chief
Colonel.—Sir D. C. Drury-Lowe, K.C.B. (Lieut.-Gen.)
Lieutenat-Colonel.—E. A. Belford
Majors.—F. W. Benson (Attached Egyptian Army)
M. G. Neeld
H. C. Jenkins
Captains.—C. H. Purvis
H .Fortescue
E. B. Herbert
Hon. L. H. D. Fortescue

Captains.—C. J. Anstruther
W. G. Renton
C. Coventry
H. W. R. Ricardo
Hon. H. A. Lawrence
Lieutenants.—G. C. C. D'Aguilar
G. F. Milner
C. A. S. Warner
F. P. M. Maryon-Wilson
B. P. Portal
N. T. Nickalls
H. M. Jessel
V. S. Sandeman
T. G. Collins
2nd Lieuts.—Prince Adolphus of Teck
H. C. Noel
W. F. Egerton
W. A. Tilney
Adjutant.—Hon. H. A. Lawrence
Riding-Master.—W. Pilley
Quartermaster.—C. Clarke (Hon. Lieutenant)

1894.

Colonel-in-Chief.
H.R.H. Duke of Cambridge, Field Marshal, Commander-in-Chief

Colonel.—Sir D. C. Drury-Lowe, K.C.B. (Lieut.-Gen.)
Lieutenant-Colonel.—E. A. Belford
Majors.—M. G. Neeld
C. H. Purvis
H. Fortescue
Captains.—E. B. Herbert
Hon. L. H. D. Fortescue
C. J. Anstruther
W. G. Renton
C. Coventry
H. W. R. Ricardo
Hon. H. A. Lawrence
Lieutenants.—G. C. C. D'Aguilar
C. A. S. Warner
B. P. Portal
N. T. Nickalls
H. M. Jessel
V. S. Sandeman
T. G. Collins
Prince Adolphus of Teck
H. C. Noel
2nd Lieutenants.—W. F. Egerton
W. A. Tilney
Sir F. Burdett, Bt.
Adjutant.—Hon. H. A. Lawrence
Riding-Master.—W. Pilley
Quartermaster.—C. Clarke

APPENDIX B

QUARTERS AND MOVEMENTS OF THE 17TH LANCERS SINCE THEIR FOUNDATION

[[1] signifies headquarters]

1759. *November 7th.*—Warrant for raising the regiment.
November 26th (?)—First rendezvous. Watford and Rickmansworth.
December.—Coventry.
1760. *October.*—Haddington,[1] Musselburgh.
1761. *August.*—Perth,[1] Falkland, Aberdour, Cupar, Culross, Leven.
1762. *June.*—Musselburgh[1] (2 troops), Dalkeith (2), Hamilton.
September.—Haddington,[1] Dalkeith, Dunbar, Hamilton, Musselburgh, Linlithgow.
1763. *January.*—Haddington[1] (2), Dalkeith, Dunbar, Musselburgh, Linlithgow.
1764 to 1771.—Ireland. [Gap in the muster-rolls; 2 troops in the Isle of Man 1766.]
1772. *January.*—Clonmel[1] (3), Clogheen (2), Leightonbridge (1).
July.—Kilkenny[1] (2), Carrick (2), Ross (2).
1773. *January.*—Kilkenny[1] (2), Carrick (2), Ross, Leightonbridge.
July.—Carlow,[1] Athy, Tullow, Callen.
1774. *January.*—Carlow,[1] Athy, Tullow, Callen.
July.—Maryborough,[1] Mount Mellick.
1775. *April.*—Embarked for Boston; arrived 10-15 June.
America, active service.
1776. *March.*—Embarked for Halifax. *June.*—Left Halifax.
July.—Landed Staten Island. *August.*—Mustered Staten Island.
1777. *January.*—Mustered at New York.
May. ,, Perth and Amboy.
August. ,, Camp, New York Island, and Bloomendale.
1778. *February.* ,, Philadelphia.
1779. *September.* ,, Flushing, Long Island (detachment to Carolina).
1780. *May.*—Mustered at Hampstead, Long Island.

Appendix B

1780. *July.*—Mustered at East Chester.
1781. *January.* „ Haarlem, N. Y., and Hampstead, L. I.
July. „ Flushing, L. I.
1782. *January.* „ Hampstead, L. I.
July. „ Fort Knyphausen.
1783. *January.* „ New York and Haarlem.
July. „ New York.
Embarked for Ireland.
1784. *January.*—Cork (on arrival).
July.—Maryborough[1] (3), Mount Mellick (3).
1785. *January.*—Maryborough,[1] Mount Mellick.
July.—Tullamore,[1] Philipstown.
1786. *January.*—Tullamore,[1] Philipstown.
July.—Longford,[1] Navan.
1787. *January.*—Athlone,[1] Mount Mellick, Navan, "Man-of-War."
July.—Castlebar,[1] Sligo, Ballinrobe.
1788.—Castlebar,[1] Sligo, Ballinrobe.
1789.—Bandon.
1790. *July.*—Kilkenny.
1791. *January.*—Kilkenny,[1] Carrick, Ross.
July.—Kilkenny.
1792. *January.*—Kilkenny.
July.—Phœnix Park.
1793. *January.*—Collon.
July.—Lisburn.
1794.—Belturbet.
1795. *May* ?—Three troops embarked for West Indies—Jamaica.
August. „ „ „ St. Domingo.
Active service.
1796.—Jamaica, Grenada, St. Domingo.
1797. *March.*—Port Royal (3 troops) ? for embarkation.
May.—Trowbridge (2 troops ? depôt).
August.—Return from West Indies. Nottingham, Trowbridge, Gloucester, Bath, Bristol.
1798.—Canterbury (detachment on active service to Ostend).
1799.—Canterbury. Two troops to Southampton.
Summer.—Swinley Camp.
Winter.—Exeter and Taunton.
1800. *Summer.*—Bagshot Heath.
Winter.—Duffield (in aid of civil power).
1801 to 1802.—Manchester,[1] Lancaster, Chester, Bolton, Preston.
1803. *May.*—Embarked for Ireland.
Tullamore,[1] Philipstown, Carlow, Kilkenny
1804.—Clonmel,[1] Tullamore, Philipstown, Carlow, Kilkenny.

1805.—Dublin.
September.—Moved to Northampton.
1806. *April.*—Brighton, Romney, Rye, Hastings.
October.—Embarked for active service in South America.
December.—Arrived in La Plata.
1807.—Active service in South America.
November.—Embarked for England.
1808. *January.*—Disembarked at Portsmouth and marched to Chichester.
February.—Embarked for East Indies.
August.—Fort William, Calcutta.
1809. *February.*—Surat. Detachment to Persia.
1810.—Surat.
1811. *December.*—Ruttapore.
1812 to 1821.—Ruttapore. Active service, detachments 1813 to 1815; whole regiment, 1816 to 1821.
1822.—Ruttapore.
1823. *January.*—Embarked for England.
May.—Arrived in England. Quarters, Chatham.
1824. *June.*—Regent's Park Barracks.
July.—Canterbury.
1825. *June.*—Regent's Park Barracks.
July.—Brighton, Chichester.
1826. *March.*—Exeter and Topsham.
1827. *January.*—Hounslow and Hampton Court.
1828. *April.*—Dundalk, Belturbet.
1829. *May.*—Dublin.
1830. *May.*—Newbridge,[1] Armagh, Navan, Kells, Kilkenny.
1831. *April.*—Limerick,[1] Ennis, Newmarket, Adair.
June.—Headquarters to Ballincollig.
1832. *April.*—Portobello Barracks, Dublin.
June.—Newport,[1] Berkeley, Dursley.
July.—Dursley,[1] Wootton-under-Edge.
November.—Headquarters to Gloucester.
(Cholera year.)
1833. *March.*—Hounslow,[1] Hampton Court, Kensington.
1834. *May.*—Leeds,[1] Burnley.
1835. *May.*—Manchester.
1836. *April.*—Norwich, Ipswich.
1837. *May.*—Coventry, Northampton.
1838. *June.*—Portobello Barracks, Dublin
1839. *January.*—Royal Barracks, Dublin.
August.—Portobello Barracks.
1840.—Portobello Barracks.
1841.—Glasgow, Edinburgh.[1]

Appendix B

1842.—Leeds.
1843. *April.*—Nottingham.[1]
[Autumn.]—Birmingham.[1]
1844. *May.*—Hounslow.[1]
1845. *April.*—Brighton.[1]
1846. *June.*—Dundalk.[1]
1847. *April.*—Island Bridge,[1] Portobello and Royal Barracks.
October.—Royal Barracks.
1848 to 1849.—Royal Barracks, Dublin.
1850. *April.*—Newbridge,[1] Clonmel, Kilkenny, Waterford, Carrick.
1851. *April.*—Woolwich.
October.—Canterbury.
1852. *June.*—Brighton,[1] Christchurch, Trowbridge.
1853. *March.*—Brighton,[1] Dorchester.
June.—Chobham.
July.—Hounslow,[1] Hampton Court.
1854. *April.*—Sailed for active service in the Crimea. Depôt, Canterbury.
1855.—Crimea.
1856. *April.*—Left the East for Ireland.
May.—Cahir Barracks,[1] Fethard, Clonmel, Clogheen, Limerick.
September.—Portobello Barracks.
1857. *March.*—Island Bridge Barracks.
October.—Embarked for active service in India. Depôt, Canterbury.
1858. *February.*—Arrived Kirkee, Bombay.
Pursuit of Tantia Topee.
1859. *May*—Gwalior.
1860. *January.*—Left Gwalior.
April.—Secunderabad.
1861 to *December* 1864.—Secunderabad.
1865. *January.*—Embarked for England.
May.—Colchester.
1866. *March.*—Aldershot.
1867. *August.*—Brighton,[1] Shorncliffe.
1868. *June.*—Woolwich,[1] Kensington, Hampton Court.
August.—Hounslow, Kensington, Hampton Court.
1869. *July.*—Edinburgh,[1] Hamilton.
1870. *April.*—Royal Barracks, Dublin.
1871. *April.*—Longford,[1] Athlone, Ballinrobe, Castlebar, Gort.
1872. *May.*—Ballincollig, Limerick, Cork, Fermoy, Clogheen.
1873. *July.*—Curragh.
August.—Island Bridge Barracks, Dublin.
1874. *August.*—Dundalk,[1] Belfast, Belturbet (1 troop in December)
1875. *June.*—Island Bridge[1] and Royal Barracks, Dublin.
1876. *June.*—Embarked for England for autumn manœuvres.

1876. *September.*—East Cavalry Barracks, Aldershot.
1877. *August.*—Leeds,[1] Preston, Sheffield.
1878. [*May.*—Detachments to Burnley, Blackburn, and Clitheroe, in aid of civil power.]
July.—Aldershot.
September.—Hounslow,[1] Hampton Court.
1879. *February.*—Embarked for active service in South Africa. Depôt, Hounslow
April.—Arrived Durban.
October.—Embarked for India.
November.—Arrived at Mhow.
1880 to *January* 1884.—Mhow. Depôt, Canterbury.
1884. *January and February.*—Lucknow.
1885 to 1890.—Lucknow.
1890. *October.*—Embarked for England.
November.—Shorncliffe (one squadron in Egypt).
1891. *July.*—Hounslow.
1892. Hounslow, Hampton Court, and Kensington.
1893. *September.*—Preston [1] [Derby, Alfreton, Normanton (in aid of civil power)] and Birmingham.
1894. Leeds,[1] Birmingham.

APPENDIX C

PAY OF ALL RANKS OF A LIGHT DRAGOON REGIMENT

1764

S. = "Subsistence." A. = Arrears. G. = Grass money.

Rank		£	s.	d.
Colonel.	S.	£483	12	6
	A.	112	13	3
		£596	5	9
Lieut.-Colonel.		£337	12	6
		79	14	9
		£417	7	3
Major.		£282	17	6
		66	7	0
		£349	4	6
Captain.		£209	17	6
		54	3	5
		£264	0	11
Capt.-Lt. & Lieut.	S.	£127	15	0
	A.	25	11	4
		£153	6	4
Cornet.		£109	10	0
		26	15	8
		£136	5	8
Chaplain.		£91	5	0
		22	6	4
		£113	11	4
Adjutant.		£82	2	6
		20	1	9
		£102	4	3
Surgeon.	S.	£82	2	6
	A.	20	1	9
		£102	4	3
Surgeon's Mate.		£54	15	0
		4	17	5
		£59	12	5
Quartermaster.		£73	0	0
		20	13	10
		£93	13	10
Sergeant.	S.	£18	5	0
	A.	9	9	0
	G.	1	11	10
		£29	5	10
Corporal.	S.	£12	2	8
	A.	6	2	0
	G.	1	11	10
		£19	16	6
Trumpeter.		£18	5	0
		7	16	0
		1	11	10
		£27	12	10
Farrier.		£9	2	0
		3	1	0
		1	11	10
		£13	14	10[1]
Light Dragoon.		£9	2	0
		3	1	0
		1	11	10
		£13	14	10

[1] Besides a halfpenny per day per horse of his troop.

1796

All the allowances hitherto known under the head of

Bread money,
Grass money,
Poundage money,
New allowances for necessaries,

to be comprised under one head, and form a daily rate of allowance. Such daily rate for non-commissioned officers and men of the cavalry (after deduction of 1s. 8d. per man for horsecloth and surcingle) to be 3½d. *per diem.*

APPENDIX D

HORSE FURNITURE AND ACCOUTREMENTS OF A LIGHT DRAGOON (WITH PRICES THEREOF) IN 1759

Item	£	s.	d.
Saddle	£1	1	0
Holsters	0	5	8
Stirrup Leather	0	1	3
Tinned Stirrups	0	3	6
Girths and Surcingle[1]	0	2	6
Crupper	0	0	11
Breastplate	0	1	2
Furniture complete with Leather Seat and Embroidery	1	7	6
Crupper Pad	0	1	3
Point Straps and Loops	0	1	0
Carbine Bucket	0	1	8
Bucket Strap	0	0	9
Carbine Strap	0	0	3½
2 long Baggage Straps	0	1	6
2 single " "	0	1	4
1 middle " Strap	0	0	6½
2 Cloak Straps	0	0	8
1 middle Cloak Strap	£0	0	3
Bridle and Bridoon	0	4	6
Tinned Bit	0	3	0
Linking Collar, brown	0	2	6
" " white	0	1	6
Pair Leathered Canvas Bags for curry comb and brushes	0	3	2
Curry Comb and Brush[1]	0	2	3
Mane Comb and Sponge[1]	0	0	8
Horse Cloth[1]	0	4	9
Snaffle Watering Bridle[1]	0	2	0
Carbine	2	0	0
Pair of Pistols	1	10	0
Sword	0	12	0
" Belt	0	5	0
Shoulder Belt	0	5	0
Cartridge Box and Belt	0	2	8

"NECESSARIES" OF A CAVALRY SOLDIER, 1795

3 Shirts
2 pairs Shoes
1 " Gaiters
2 " Stockings
Forage Cap
Saddle Bag
1 pair Canvas or Woollen Overhose
1 Stock
1 Black Ball
1 Canvas or Woollen Frock or Jacket
2 Brushes
1 Curry Comb and Brush
1 Mane Comb and Sponge
1 Horse Picker

[1] Articles marked ([1]) were found at the Dragoon's expense out of his arrears and grass money. Also the following articles (besides the clothing specified in Appendix E): Goatskin holster top at 1s. 6d.; Horse picker and turnscrew, 2d.; Pair of saddle bags.

APPENDIX E

CLOTHING, ETC., OF A LIGHT DRAGOON, 1764

Coat, waistcoat, breeches, and cloak found by the Colonel by contract.

Item	£	s.	d.
Helmet	£0	16	0
Boots and Spurs	1	3	0
Watering Cap	0	2	6
4 Shirts[1] at 6s. 10d.	1	7	4
4 pairs Stockings[1] at 2s. 10d.	0	11	4
1 pair Boot Stockings	0	2	0
2 pairs Shoes at 6s.[1]	0	12	0
1 Black Stock[1]	0	0	8
1 „ „ Buckle[1]	0	0	6
1 pair Leather Breeches[1]	1	5	0
1 pair Knee Buckles[1]	0	0	8
1 pair Short Black Gaiters[1]	0	7	4
White Jacket[1 2]	0	8	6
Stable Frock	0	4	8
Pick-wire and Pan Brush	0	0	2
Worm and Oil Bottle	...		
Necessary Bags	£0	7	3
Corn Bag	0	2	6
Black Ball[1]	0	1	0
3 Shoe Brushes[1]	0	1	3
Hair Comb	0	0	6
Burnisher	0	0	6
White Portmanteau	0	8	0
1 pair of Gloves	0	1	6
Farrier's Cap	0	14	0
„ Budgets	0	14	0
„ Apron	0	1	8
„ Axe and Case	0	5	0
„ Saw and Case	0	8	6
Trumpeter's Hat and Feather	1	0	0
Trumpet	2	2	0
Sling and Tassels of crimson and white	0	10	0

All articles marked (1) supplied, according to King's regulation and custom, out of the Light Dragoon's arrears and grass money.

2 White Jacket added to the kit by the special request of the men themselves at the close of the Seven Years' War.

APPENDIX F

EVOLUTIONS REQUIRED AT THE INSPECTION OF A REGIMENT

1759

The squadron was drawn up in three ranks at open order, *i.e.* with a distance equal to half the front of the squadron between ranks.

Each squadron was told off into half-ranks, one-third of ranks, and fours.

Officers take your posts of exercise.—The officers rode out from their posts till eight or ten paces in rear of the C.O., then turned about and faced their squadrons.

Half-ranks to the right; double your files.—The left half-ranks of each squadron reined back to the half-distance between ranks, and passaged to the right until the right half-ranks were covered.

Half-ranks that doubled; as you were.—The left half-ranks passaged to the left and rode back to their original places.

(The same manœuvre then executed to the left.)

Rear ranks to the right; double your front.—The rear ranks wheeled into column of half-ranks, then wheeled (as a column) to the left and came up, the leading half-rank on the right flank of the front, and the rear half-rank on the right flank of the centre rank.

Rear ranks that doubled; as you were.—The columns of half-ranks wheeled to the right, and countermarched to their original places.

(The same manœuvre then repeated to the left.)

By two divisions to right and left about, outward, march.—Each rank of each squadron divided in the centre, and wheeled, the right half-ranks to right about, and the left half-ranks to left about; whereby each squadron was formed into two divisions, with an interval between them, facing to the rear.

Wheel to the right and left about to your proper front.—The original formation resumed.

Centre rear ranks move up to your order.—"Order" allowed a distance equal to one-third of the squadron's frontage between ranks.

By three divisions wheel to the right.—We should now give the word "Divisions, right wheel."

To the right.

To the right about.

(Same manœuvre repeated to the left.)

Centre and rear ranks move forward to your close order.—Close order reduced the distance between ranks to the space required for four men to wheel abreast.

By fours wheel to the right about.

By fours wheel to the left about.

Officers take post in front of your squadrons.

Squadrons wheel to the right; march.

To the right.

To the right about.

The same then was repeated to the left; and the evolutions came to an end, the trumpets blowing a march till the inspecting officer was out of sight.

THE END

www.ingramcontent.com/pod-product-compliance
Ingram Content Group UK Ltd.
Pitfield, Milton Keynes, MK11 3LW, UK
UKHW021840270726
14058UKWH00002B/254

9 781843 421207